SPINOZA'S GEOMETRY OF POWER

This work examines the unique way in which Benedict de Spinoza (1632–77) combines two significant philosophical principles: that real existence requires causal power, and that geometrical objects display exceptionally clearly how things have properties in virtue of their essences. Valtteri Viljanen argues that underlying Spinoza's psychology and ethics is a compelling metaphysical theory, according to which each and every genuine thing is an entity of power endowed with an internal structure akin to that of geometrical objects. This allows Spinoza to offer a theory of existence and of action – human and non-human alike – as dynamic striving that takes place with the same kind of necessity and intelligibility that pertain to geometry. Viljanen's fresh and original study will interest a wide range of readers in Spinoza studies and in early modern philosophy more generally.

VALTTERI VILJANEN is a postdoctoral fellow at the University of Turku, Finland. He is co-editor (with Juhani Pietarinen) of the anthology *The World as Active Power: Studies in the History of European Reason* (2009), and the author of a number of journal articles on Spinoza's philosophy.

SPINOZA'S GEOMETRY OF POWER

VALTTERI VILJANEN

CAMBRIDGE
UNIVERSITY PRESS

CAMBRIDGE
UNIVERSITY PRESS

University Printing House, Cambridge CB2 8BS, United Kingdom

Cambridge University Press is part of the University of Cambridge.

It furthers the University's mission by disseminating knowledge in the pursuit of education, learning and research at the highest international levels of excellence.

www.cambridge.org
Information on this title: www.cambridge.org/9781107007802

First published 2011

A catalogue record for this publication is available from the British Library

Library of Congress Cataloguing in Publication data
Viljanen, Valtteri.
Spinoza's Geometry of Power / Valtteri Viljanen.
p. cm.
Includes bibliographical references.
ISBN 978-1-107-00780-2
1. Spinoza, Benedictus de, 1632–1677. 2. Ontology. 3. Causation. 4. Power (Philosophy)
5. Geometry – Miscellanea. I. Title.
B3998.V55 2011
199′.492–dc22
2011007624

ISBN 978-1-107-00780-2 Hardback
ISBN 978-1-107-41738-0 Paperback

Contents

Acknowledgements

This monograph is the main result of more than ten years of work. The journey really began in 1998, when Juhani Pietarinen gave a seminar lasting a full academic year on Spinoza's *Ethics*. That seminar was crucial for arousing my interest in early modern philosophy in general and in Spinoza's thought in particular. Since that time, Juhani has not only consistently encouraged me to interpret Spinoza from a dynamistic point of view but also shown me the excitement of tracking large-scale lines of development in the history of philosophy. My good fortune did not end with the fading millennium: when, a few years later, I began my Ph.D. studies, Olli Koistinen was appointed as my supervisor. His competence and unfailing support, which I have had the great privilege to enjoy over many years and in innumerable discussions, have had a profound impact on my philosophical life. To these two exemplary mentors I can only express my deepest gratitude. The present work, much revised, has grown out of my dissertation.

I also wish to thank all my colleagues at the Department of Philosophy of my *alma mater*, the University of Turku, for creating such an excellent working environment. Arto Repo's role has been particularly important: for me, his exceptional kindness and helpfulness have been no less important than his philosophical acuity. In the congenial atmosphere of our traditional weekly research seminar, the so-called 'Rationalist Circle', I have been allowed to try out new ideas. I am grateful to all those who have, over the years, taken part and shared their views in the Circle, especially (in addition to those already mentioned) Markku Keinänen, Tapio Korte, and Hemmo Laiho.

The Finnish philosophical community outside my home town has been important for me as well. I have found particularly inspiring and heart-warming the companionship and support of a group of young historians of philosophy: Jani Hakkarainen, Toni Kannisto, Juhana Lemetti, Vili Lähteenmäki, Ville Paukkonen, Mika Perälä, and Markku Roinila. I would also like to thank the audiences at the universities of Helsinki, Jyväskylä, and Tampere, in which I have had, on several occasions, the opportunity to present material that has found its way into this work.

My prolonged visits to Tübingen (2002–3), Paris (2004), and Uppsala (2008) have enabled me to discuss my work with many people, and I would particularly like to thank the following colleagues: Lilli Alanen, Tomas Ekenberg, Chantal Jaquet, Minna Koivuniemi, Pierre-François Moreau, Paulina Remes, and Andreas Schmidt. A very special thanks goes to Peter Myrdal and Erik Åkerlund for many rewarding discussions on parts of this manuscript.

This work has benefited from a considerable number of expert comments provided by the worldwide community of scholars, and I would like to thank all those with whom I have discussed my views in seminars, workshops, and conferences. I am especially grateful to Steven Barbone and Don Garrett for their insightful reports on the dissertation, and to Don also for all the support thereafter; to John Carriero, Jon Miller, and Lee Rice for instructive feedback on some earlier material eventually incorporated into this work; to Charles Jarrett for his interest in and constructive comments on my work throughout its many stages; to Mogens Lærke for providing a wealth of detailed and helpful comments on a late version of this manuscript; and to Robert Pasnau and Don Rutherford for their valuable suggestions on the Introduction – Don also deserves a special thanks for a memorable discussion in Kilpisjärvi that settled the title of this work.

My warmest thanks are also due to the people at Cambridge University Press, especially to the two anonymous referees for their extremely helpful reports, and to my editor, Hilary Gaskin, for her expertise and effort in bringing the process to completion. I would also like to thank Jo North for her diligent copy-editing.

Last but definitely not least, I am grateful to my family and especially to my wife, friend, and kindred philosophical spirit Hanna Meretoja, whose commitment to academic life has always set an example for me. Without her affection and encouragement the research leading to this book would hardly even have begun; so it is to her that I dedicate this work.

This monograph contains, in a revised form, material that has previously been published in the following articles:

'Field Metaphysic, Power, and Individuation in Spinoza'. *The Canadian Journal of Philosophy* 37 (3), 2007: 393–418.

'On the Derivation and Meaning of Spinoza's *Conatus* Doctrine'. In *Oxford Studies in Early Modern Philosophy, Volume* IV, ed. Daniel

Garber and Steven Nadler, pp. 89–112. Oxford: Clarendon Press, 2008.

'Spinoza's Essentialist Model of Causation'. *Inquiry* 51 (4), 2008: 412–37.

Permission to incorporate this material into the present work is gratefully acknowledged.

Finally, I would like to acknowledge that the work on this monograph has been financially supported by the Academy of Finland (project number 127410).

Abbreviations

SPINOZA'S WORKS

Collected works

C *The Collected Works of Spinoza, Volume 1*. Translated and edited by Edwin Curley. Princeton University Press, 1985.

G *Spinoza Opera* I–IV. Edited by Carl Gebhardt. Heidelberg: Carl Winter, 1925.

S *Complete Works*. Translated by Samuel Shirley, edited by Michael L. Morgan. Indianapolis: Hackett, 2002.

The Ethics

I am using Curley's translation (in C). Unless otherwise indicated, a reference to a work by Spinoza is to the *Ethics*; the first arabic number specifies part of the work and the abbreviations following it are as follows:

a	axiom
defaff	definition of the affects (in the third part of the *Ethics*)
app	appendix
c	corollary
d	definition (when not after a proposition number)
d	demonstration (when after a proposition number)
le	lemma
p	proposition
po	postulate
pr	preface
s	scholium

For instance, 1p8s2 refers to the second scholium of the eighth proposition in the first part of the *Ethics*.

Other works by Spinoza

CM	*Metaphysical Thoughts* (*Cogitata Metaphysica*)
Ep	*Correspondence* (*Epistolae*)
KV	*Short Treatise on God, Man and His Well-being* (*Korte Verhandeling van God, de Mensch, en des zelfs Welstand*)
PPC	*Descartes' 'Principles of Philosophy'* (*Principia Philosophiae Cartesianae*)
TdIE	*Treatise on the Emendation of the Intellect* (*Tractatus de Intellectus Emendatione*)
TP	*Political Treatise* (*Tractatus Politicus*)
TTP	*Theological-Political Treatise* (*Tractatus Theologico-Politicus*)

Other sources

Works by all other authors up to and including Friedrich Nietzsche are referred to by the abbreviations in the following list. Works by later authors are listed, with their publication dates, in the bibliography at the end of this volume.

AG	Leibniz, Gottfried Wilhelm. *Philosophical Essays.* Edited and translated by Roger Ariew and Daniel Garber. Indianapolis: Hackett, 1989.
BGE	Nietzsche, Friedrich. *Beyond Good and Evil: Prelude to a Philosophy of the Future.* Edited by Rolf-Peter Horstmann and Judith Norman, translated by Judith Norman. Cambridge University Press, 2001.
CAM	Thomas Aquinas. *Commentary on Aristotle's* Metaphysics. Translated by John P. Rowan. Notre Dame: Dumb Ox Books, 1995.
CSM	Descartes, René. *The Philosophical Writings of Descartes* I–II. Translated by John Cottingham, Robert Stoothoff, and Dugald Murdoch. Cambridge University Press, 1985.
CSMK	Descartes, René. *The Philosophical Writings of Descartes* III. *The Correspondence.* Translated by John Cottingham, Robert Stoothoff, Dugald Murdoch, and Anthony Kenny. Cambridge University Press, 1991.
CWA	Aristotle. *The Complete Works of Aristotle. The Revised Oxford Translation* I–II. Edited by Jonathan Barnes. Princeton University Press, 1984.
Duties	Cicero. *On Duties.* Edited by M. T. Griffin and E. M. Atkins, translated by E. M. Atkins. Cambridge University Press, 1991.
EG	Galilei, Galileo. *The Essential Galileo.* Edited and translated by Maurice A. Finocchiaro. Indianapolis: Hackett, 2008.

EW	Hobbes, Thomas. *The English Works of Thomas Hobbes* I–XI. Edited by William Molesworth. London: John Bohn, 1839–45.
LM	Kant, Immanuel. *Lectures on Metaphysics*. Translated and edited by Karl Ameriks and Steve Naragon. Cambridge University Press, 1997.
MD	Suárez, Francisco. *On Efficient Causality. Metaphysical Disputations 17, 18, and 19*. Translated by Alfred J. Freddoso. New Haven: Yale University Press, 1994.
ME	Cicero. *On Moral Ends*. Edited by Julia Annas, translated by Raphael Woolf. Cambridge: Cambridge University Press, 2001.
NG	Cicero. *The Nature of the Gods*. Translated by P. G. Walsh. Oxford University Press, 1998.
NM	Kant, Immanuel. *Attempt to Introduce the Concept of Negative Magnitudes into Philosophy*. In *Theoretical Philosophy, 1755–1770*. Translated and edited by David Walford in collaboration with Ralf Meerbote. Cambridge University Press, 1992.
PW	Newton, Isaac. *Philosophical Writings*. Edited by Andrew Janiak. Cambridge University Press, 2004.
SCG	Thomas Aquinas. *The Summa Contra Gentiles*. Translated by the English Dominican Fathers. London: Burns, Oates, and Washbourne, 1934.
Sophist	Plato. *Sophist*. In *The Collected Dialogues of Plato*. Edited by Edith Hamilton and Huntington Cairns. Princeton University Press, 1989.
ST	Thomas Aquinas. *Summa Theologica*. Translated by the Fathers of the English Dominican Province. New York: Christian Classics, 1981.
SW	Thomas Aquinas. *Selected Writings of St. Thomas Aquinas*. Translated by Robert P. Goodwin. Upper Saddle River: Prentice Hall, 1965.
World	Descartes, René. *The World*. In *The World and Other Writings*. Translated and edited by Stephen Gaukroger. Cambridge University Press, 1998.
WP	Nietzsche, Friedrich. *The Will to Power*. Translated by Walter Kaufmann and R. J. Hollingdale, edited by Walter Kaufmann. New York: Vintage Books, 1968.

Introduction

We must begin with two convictions prevalent in Western thought and old as philosophy itself, one that concerns us as cognitive beings, another the nature of reality itself. To take the better-known first: in geometry, the scientifically oriented human mind has found its apogee of clarity, lucidity, and certainty. Through the tumults that gave birth to the modern era, the status of geometry remained strong: nature itself was seen, to an increasing degree, in its terms. There is perhaps no more striking indication of this than the memorable passage in Galileo's *The Assayer* of 1623:

> Philosophy is written in this all-encompassing book that is constantly open before our eyes, that is the universe; but it cannot be understood unless one first learns to understand the language and knows the characters in which it is written. *It is written in mathematical language, and its characters are triangles, circles, and other geometrical figures; without these it is humanly impossible to understand a word of it,* and one wanders around pointlessly in a dark labyrinth. (*EG*, p. 183, emphasis added)

Only geometry is thus able to capture the very texture of nature: the universe is written with its characters. This spirit finds its way to the thought of Benedict de Spinoza (1632–77), who in his masterpiece, the *Ethics*, claims that however profound and intricate a question, it can be answered, but only after realizing a fundamental truth:

> So they [men] maintained it as certain that the judgments of the Gods far surpass man's grasp. This alone, of course, would have caused the truth to be hidden from the human race to eternity, if Mathematics, which is concerned not with ends, but *only with the essences and properties of figures,* had not shown men another standard of truth. (1app; C, p. 441; G II, p. 79, emphasis added)

That the philosopher renowned for expressing his ideas in geometrical order has a high regard for mathematics is of course only what we should expect; but it should be appreciated that here geometry is *not* conceived of as a method or manner of exposition. Rather, it is the branch that has revealed *the inner make-up of things.* This idea manifests itself in various ways in Spinoza's thought, and forms one of the recurring themes of this study.

The other conviction, less often acknowledged but still extremely influential, finds its expression in Plato's *Sophist*, whose Eleatic Stranger suggests that

> anything has real being that is so constituted as to possess any sort of power [*dunamin*] either to affect anything else or to be affected, in however small a degree, by the most insignificant agent, though it be only once. *I am proposing as a mark to distinguish real things that they are nothing but power* [*dunamis*]. (*Sophist* 247d–e, emphasis added)[1]

In other words, real or actual existence requires causal power; having no effects implies non-existence. Much of the ensuing dominant Aristotelian-scholastic metaphysics takes the linkage between existence and power as given: no natural agent is without its causal powers. Thus it should not be particularly surprising that in Spinoza we find a thinker who not only equates God-or-Nature's essence with power (1p34) but one who argues that as expressions of God's power, the innermost nature of temporal existents is striving power – *conatus* – to persevere in being (3p6–p7).

That geometry is important for Spinoza and that the doctrine of finite things as striving entities is the basis of his theory of human temporal existence are, I take it, beyond controversy. But the unique way in which Spinoza combines the two traditional convictions results in something extremely significant, and something that forms the subject matter of this study, namely, an ontology on which is based a specific view of human individuality and agency. I thus want to develop a unifying overall interpretation of Spinoza's metaphysical thought with a definite centre of gravity: the idea that on the pages of Spinoza's main work and underpinning his ethics proper there is – deeply integrated but still by no means concealed – what may be called *a geometry of power.* The epithet is meant to capture what this study defends above all: *each and every genuine thing is an entity of power endowed with an internal structure akin to that of geometrical objects.* Based on this, Spinoza aspires to offer us a theory of existence – human and non-human alike – as a dynamic affair, but one that takes place with the same kind of necessity and intelligibility that pertain to geometry. It is a sign of challenges to come in advancing an approach of this kind that the two traditional convictions seem so uneasily combinable; after all, neither power nor change pertains to geometrical objects. But I believe it is the interpretative path to be taken, and one that leads us to a philosophy designed to revolutionize not only the

[1] Translation by F. M. Cornford.

view we have of the world as a whole but also our understanding of human existence and happiness.

As suggested by the passage of *The Assayer*, Spinoza's philosophical revolution has its roots in the upheavals of the natural sciences. The rupture with the tradition is considerable: gone are genera and species, substantial and accidental forms, different kinds of change and so on; instead, natural things are seen as mathematical entities concretized – impenetrable and in motion, but nevertheless geometrical figures by their very nature. Given the new 'mathematicized' view's economy and the ease of understanding the world it promised, it is not particularly difficult see why a mind yearning for a systematic grasp of things would find it attractive. Two of Spinoza's immediate predecessors, Descartes and Hobbes, certainly felt that attraction; and turning to Spinoza's younger contemporaries, consider the following statement by Newton in *De Gravitatione*:

> [S]paces are everywhere contiguous to spaces, and extension is everywhere placed next to extension, and so there are everywhere common boundaries of contiguous parts; that is, there are everywhere surfaces acting as boundaries to solids on this side and that; and everywhere lines in which parts of the surfaces touch each other; and everywhere points in which the continuous parts of lines are joined together. And hence *there are everywhere all kinds of figures, everywhere spheres, cubes, triangles, straight lines, everywhere circular, elliptical, parabolical, and all other kinds of figures*, and those of all shapes and sizes, even though they are not disclosed to sight. (*PW*, p. 22, emphasis added)[2]

But Spinoza takes the crucial step further: he is convinced that everything, thus *also the realm of the mental*, is endowed with *the precisely same kind of structure and intelligibility* we can find in the material world.

The very first step in discerning the metaphysical basis of Spinoza's theory of human existence is to examine this general view of the structure of the nature of things. The fact that in the *Ethics* Spinoza leaves many of these basic aspects of his thought unexplicated increases the difficulty involved in carrying out this task. However, my goal is to show that a careful reading of his earlier work such as *Treatise on the Emendation of the Intellect* and *Metaphysical Thoughts* can shed crucial light on many better-known tenets of Spinoza's

[2] The date of this posthumously published manuscript is a matter of controversy. It has been suggested that the text would originate already from the 1660s; for Andrew Janiak's survey of the different views on the issue, see *PW*, p. xviii. Mordechai Feingold (2004, p. 194) claims that 'the document in its present form is of late composition, albeit incorporating earlier material' and estimates that the work's 'origins might be more precisely dated to around 1671, and to a course of lectures Newton delivered at Cambridge against Descartes' mechanics and Henry More's hydrostatics' (p. 26). So, interestingly, it seems to have been a work in progress when Spinoza was writing his *Ethics*.

masterpiece. Here I will approach Spinoza not so much as a naturalist influenced by the advances in the mechanical sciences but as a rationalist metaphysician inspired by geometry. In fact, the opening chapter of this study argues that from early on, Spinoza endorses and develops a general theory of *the being of essences* which aims at explicating the very factors that determine the form that existence takes, and ends with the conviction that all genuine things are endowed with an internal structure not unlike the one we find in geometrical objects. The result is a view that underpins the all-important theory of the two aspects of reality, eternal and temporal, of the *Ethics*. I argue that in Spinoza's rationalism, the natures of things and what takes place in temporality are determined from eternity.

Discerning Spinoza's theory on the foundational but intangible ontological features of the world allows us to begin the journey towards a theory concerning actual 'flesh and blood' entities of temporal existence to which pertain dynamics in the most salient sense. As noted, real existence is, traditionally as well as for Spinoza, causally efficacious existence (this in fact is what makes the introduction of the concept of power – a causal notion – legitimate and useful). An analysis of causation is thus required of any overall reading of Spinoza's metaphysics, and especially of one that underscores the dynamism of his system. As Chapter 2 shows, it is precisely here – importantly for my main thesis – that we find Spinoza's geometrical tendencies at work: his doctrine of causation is *derived from* the geometry-inspired doctrine of being. A particular conception of essentialism is the philosophical centre of gravity: just as the structure of being is determined by the eternal natures of things, effects follow from the essences of things.

With all this in place, I turn to the focal concept of this study, that of power. Chapter 3 starts with a contextual discussion explicating how the concept operated and came to be questioned in metaphysics before Spinoza; this together with the already acquired understanding of the essentialist model of causation allows us to discern the concept's role in Spinoza's thought as well as his reasons for including it in his system. Clearly, Spinoza thinks that within his metaphysical framework, the concept of power can be assigned not only a transparent meaning but also a proper task in a theory concerning the ethically relevant features of our causal status. From the opening part of the *Ethics*, Spinoza makes the distinction between power *to exist* and power *to act*; accordingly, Chapter 3 offers an analysis of power, existence, and activity. First, it is shown how power figures in the existence of all the different types of Spinozistic entities: substances, modes, and attributes. After this, an examination of *power of acting* reveals it to be a particularly prominent feature of the causal machinery that later proves to be constantly at work in the vicissitudes of finite temporal existence. Moreover, given that the chapter shows things to be endowed with power (whether to exist or to act) in virtue of their essences, it is warranted to speak of Spinoza's *dynamic essentialism.*

Elaborating a workable metaphysic is of course a valuable undertaking in its own right, but Spinoza wants to accomplish something more with it: an adequate account of actual (i.e. temporal) human existence. In it, the notion of *conatus* emerges as the key feature: it is through the doctrine of our essential *power as striving to persevere in being* that Spinoza's geometry of power becomes a fully developed doctrine of human existence. The scientific revolution and its groundbreaking conception of motion have an impact on how Spinoza conceives the essential power of things to manifest itself in actual existence, and Chapter 4 starts by discussing the historical roots of the *conatus* doctrine as well as its place in Spinoza's metaphysical framework. The derivation of the doctrine has been severely criticized, and the apparent errors in it have been claimed to have devastating effects on the overall cogency of Spinoza's system. The main task of this chapter is to take up this issue and show that provided that Spinoza's argument is carefully reconstructed by taking properly into account his geometry-laden theory of power, the reasoning in it turns out to be, in fact, basically valid.

The reconstruction of Spinoza's argument – if not the ontological considerations of Chapter 1 – makes it increasingly clear that we should rethink what Spinoza means by the 'being' in which we strive to persevere. This leads to the subsequent discussion of the nature of the *conatus* doctrine, and Chapter 5 opens by outlining the two main interpretative positions prevailing in the scholarship, the inertial and the teleological. Despite the indisputable merits of these two interpretative traditions, it is argued that Spinoza's views on these matters have not thus far been satisfactorily discerned. The rest of the chapter elaborates a new interpretative stand along the lines of geometrical dynamism; on this view, *what we strive for is the unhindered realization of our essence – to produce being as determined by what is innermost to us.* This, in turn, requires reconsidering Spinoza's view of human agency: it is a theory according to which our actions have directions without ends.

The final chapter draws on the preceding analyses to offer a fuller picture of Spinoza's thought about human existence, or what I would call his geometrical dynamics of individuality. What individuates any finite individual – makes it the individual it is – is a specific kind of power: a striving or a resilient disposition to bring about effects derivable from a definition expressing a unique eternal essence. As long as there is such a striving, even as weak or severely opposed, the individual persists in actuality. Our limitedness implies constant passivity, and it is a central part of Spinoza's philosophical psychology to map out the most relevant ways in which human beings are passively constituted. Spinoza firmly believes a project that endeavours to provide a veritable geometry of emotions to be feasible; this calls for an examination revealing his grounds for thinking that any entity with a specific nature is determined, in particular circumstances, to everything it also passively does with the kind of necessity characteristic of geometrical objects. Finally, to offer a more comprehensive interpretation, the chapter rounds off by presenting some of the major attribute-specific ramifications of Spinoza's dynamistic metaphysics: the attribute of extension can be seen as a continuous spatial field of power in which there are particular bodies as patterns of intensifications; under the attribute of thought, ideas are formed by power of affirmation – a power which, when used freely, can only lead to understanding, that is, to forming adequate ideas.

Methodologically my approach can be described as follows. With respect to each aspect of Spinoza's geometry of power, I will attempt to find its proper historical context, situate it in that context, and then, in light of this understanding and of my understanding of what Spinoza elsewhere says, engage in detailed conceptual analysis to find out how Spinoza's system works. I know of no good reason not to see contextualization and conceptual analysis as mutually supportive: being familiar with the relevant context is often a precondition for knowing what an author is talking about and practically always useful for discerning the meaning of the terms he or she uses; successful analyses, in turn, will result in a better understanding of the philosophical context which is partly constituted by the author's works. I believe that this kind of enquiry can offer us new insights into pertinent philosophical questions.

In Spinoza's case, taking a look at Descartes's philosophy is almost always in order; with regard to some questions, it is also enough, but usually one has to dig more deeply and explicate how, for example, Hobbes or late scholastics viewed things. Of Spinoza's writings I will focus on the *Ethics*, but whenever I think an earlier work or a letter is helpful, as is especially the case with regard to some metaphysical issues discussed in the beginning, I will take it into account. Fortunately, as already mentioned, Spinoza's thought is and has been the subject of lively discussion during the last decades, and I aim to make my study as well informed as possible, doing my best to bring forward the secondary literature relevant to each topic. There is no scholar whose own time and philosophical heritage would not have a major impact on his or her approach and the sort of questions he or she considers worth asking; for me, the Anglo-American early modern scholarship flourishing today forms the most immediate background against which, and in dialogue with, I develop my own views; but I will also take the French Spinoza scholars into account and strive to make connections between them and the writers belonging to the Anglo-American tradition.

The metaphysical nature of the discussed topics, Spinoza's rationalistic approach, and my endeavour to discern, through my analysis, systematic linkages between the concepts Spinoza employs may, at times, lead to passages of a rather abstract nature. Thus, to make my text more accessible I will illustrate the discussed issues with some concrete examples and analogies to offer us a firmer grasp of them. However, here the revisionary nature of Spinoza's philosophy should be kept in mind: he would insist that if sound philosophical principles and proofs so demand, we should be ready to revise our beliefs, even those widely held to be most natural (for instance that there are innumerably many substances, or that our will is free). It seems to me that we should give any piece of philosophy a fair chance of convincing us, and this applies in particular to texts from which we have a considerable amount of historical distance, such as the *Ethics*, strange as they often first appear to our contemporary eye. Otherwise we run the risk of dogmatically demoting them to the status of historical curiosities, which would keep us from learning from them. And indeed, I believe that the dynamistic view of the nature of our individuality, agency, and happiness that this study aims at explicating is highly compelling, and merits serious attention.

CHAPTER I

Spinoza on being

No doubt, the spirit of the early modern science had a great impact on Spinoza's cast of mind. But the pre-eminent aspect of this influence is *not* the one the victories of the mechanical sciences had on the conception of causation and of the nature of bodies; rather, it is the effect the new mathematical understanding of the realm of nature had on Spinoza's deepest assumptions concerning the way in which all genuine things are built, or how the basic nature of their being is determined. A clear grasp of these commitments will later prove vital for understanding Spinoza's way of thinking about power and related topics.

It is not only in the *Ethics* that Spinoza tackles fundamental ontological questions; we have much to gain from his early works. *Metaphysical Thoughts* (published in 1663) is an especially relevant tract for our purposes, for its first part discusses the question of being and the way it should be classified. The work is an appendix to a presentation of Descartes's *Principles of Philosophy* in the geometrical manner, and it is not always completely clear to what extent Spinoza is voicing his own views; but I believe that especially the first two chapters of the opening part of the *Metaphysical Thoughts* deserve to be read very carefully. The other two early works I will take up are the *Treatise on the Emendation of the Intellect* (c. 1661) and the *Short Treatise on God, Man, and His Well-Being* (c. 1662, discovered in the nineteenth century). In the latter section of this chapter, I will discuss the way in which the ideas expressed in these works are further elaborated in the *Ethics*.

EARLY SPINOZA ON THE BEING OF ESSENCES

Metaphysical Thoughts begins with a warning: we should keep firmly in mind that beings of reason, that is, modes of thinking which help 'us to more easily *retain*, *explain*, *and imagine* the things we have understood' (*CM* 1.1; C, p. 300; G 1, p. 233), should never be taken for real beings, or as referring to such. To this category of entities belong such cognitively generated things as genera and species as well as ideas used to measure duration, such as time. Having emphasized this, Spinoza moves on to discuss different types of being. The discussion is fairly convoluted, but it contains a distinction made *within* the realm of reality, or real being, whose importance cannot be overstated: the distinction between *the being of essence* and *the being of existence*. The crucial passage reads as follows:

Hence we can easily reply to the questions that are usually raised concerning essence. These questions are as follows: [Q1] whether essence is distinguished from existence? and [Q2] if it is distinguished, whether it is anything different from the idea? and [Q3] if it is something different from an idea, whether it has any being outside the intellect? [A3] The last of these must surely be granted.

[A1] To the first question we reply by making a distinction: in God essence is not distinguished from existence, since his essence cannot be conceived without existence; but in other things it does differ from and certainly can be conceived without existence. [A2] To the second we say that a thing that is conceived clearly and distinctly, *or*[1] truly, outside the intellect is something different from the idea.

But again it is asked [Q4] whether that being outside the intellect is by itself or has been created by God. [A4] To this we reply that the formal essence neither is by itself nor has been created, for both these presuppose that the thing actually exists. Rather it depends on the divine essence alone, in which all things are contained. So in this sense we agree with those who say that the essences of things are eternal. (*CM* 1.2; C, pp. 304–5; G 1, pp. 238–9)[2]

Slightly later Spinoza adds,

although the essences of nonexistent modes are comprehended in their substances, and their *being of essence* [*esse essentiae*] is in their substances, nevertheless we wished to recur to God in order to explain generally the essence of modes and of substances, and also because the essence of modes has only been in their substances after the creation of the substances and we were seeking the eternal *being of essences*. (*CM* 1.2; C, p. 305; G 1, p. 239)[3]

If these passages sound opaque, it is at least partly because they contain so incredibly much; at this point I would like to highlight the following points. First, there is the standard medieval distinction between essence and

1 Edwin Curley italicizes the word 'or' when translating *sive* or *seu*, which generally denote equivalence, not alternative (C, p. xv).

2 I have inserted the square brackets with abbreviations to help in structuring the discussion.

3 I think it can be safely assumed that in these passages, Spinoza is expressing his own views – not only because of the occurrence of such locutions as 'we reply', 'we agree', and 'we wished to recur' but also because the passages are in line with what he writes elsewhere. I think Spinoza is here formulating certain enduring ideas of his in a terminology somewhat different from the one adopted in the *Ethics*, and in a way that is of considerable help in understanding the masterpiece itself.

existence. Second, and more radically, essences are something which have an ontological status of their own: they are endowed with being independent from that of actually existing things (save God) as well as from that of ideas. In other words, essences form a category of entities which is not ideational in the sense that its inhabitants would exist only in our ideas, as their objects (they have being 'outside our intellect'). Where, then, are those entities located? The answer is that, third, the being of essences is something contained or comprehended in God, for it depends 'on the divine essence alone'. Fourth, the essences in question are eternal, completely outside time.

Before unpacking these contentions and the presuppositions they carry, we should stop and ask, why all this talk of essences and their being? What is the purpose of introducing such entities? Spinoza's answer – which is in consonance with the tradition – to this question remains unchanged throughout his philosophical career, and its basic idea finds expression already in another work dating from the same period as the *Metaphysical Thoughts*. The opening chapter of the *Short Treatise* strongly suggests that a *definite nature* (i.e. a definite essence) is what individuates its possessor, for Spinoza says that it is that '*by which the thing is what it is*, and which cannot in any way be taken from it without destroying it' (*KV* 1.1; C, p. 61; G 1, p. 15, emphasis added).[4] For instance, to have a valley belongs to the essence of a mountain, and as we have seen, Spinoza is keen to emphasize the atemporal character of these individuating factors: 'The essences of things are from all eternity and will remain immutable to all eternity', as there can never be a mountain without a valley, and '[t]his is truly eternal and immutable, and must always be in the concept of a mountain, even if it does not exist, and never did' (*KV* 1.1; C, p. 61; G 1, p. 15). Strikingly, Spinoza appears to have been consistently committed to *individual* essences; the most straightforward piece of textual evidence for this is the definition of essence in the *Ethics*, which is uneasily read in any other way:

> I say that to the essence of any thing belongs that which, being given, the thing is [NS: also][5] necessarily posited and which, being taken away, the thing is necessarily [NS: also] taken away; or that without which the thing can neither be nor be conceived, and which can neither be nor be conceived without the thing. (2d2)[6]

[4] The quotation is from a marginal note of the *Short Treatise*; I would agree with Gebhardt and Curley (see C, p. 47) that there is no reason to doubt that it is from Spinoza's pen.

[5] NS refers to the Dutch edition of Spinoza's posthumous works, *De Nagelate Schriften*; see n. 36 below.

[6] Although arguments against individual essences have been presented, I am still occupying a mainstream position here: as Christopher Martin (2008, p. 490) estimates in a recent paper that focuses on this topic, '[t]here is a strong consensus among commentators that Spinoza understands the essence of each mode to be unique to it'. Martin (2008, pp. 490–2) correctly regards 2d2 as the most explicit piece of evidence for this interpretative position.

So, to recapitulate, any true thing is endowed with an essence of its own that has a being different from and independent of the thing's actual existence but still not merely ideal in character (i.e. it is not a mere figment of our imagination), and young Spinoza calls this the being of essence(s). This essential and atemporal form of being is in an important sense real, and as it is precisely this kind of actuality-independent being that makes things what they are, it in fact is – as will become clearer below – the ontologically pre-eminent sense of being. *The being of essences is the prime layer of reality itself.*

Given the aforesaid, it is no surprise that the seminal epistemological work of the early 1660s, *Treatise on the Emendation of the Intellect*, so decidedly focuses on essences as something via which we can obtain the proper cognitive foothold on reality.[7] In that work, Spinoza argues that 'there is something real in ideas, through which the true are distinguished from the false' (§ 70; C, p. 31). What is more, the nature of this reality is something *intrinsic* to the thing of which true ideas are formed, for – and completely in keeping with the doctrine of the being of essences examined above – it does not depend on whether the thing actually exists or not. The following illustration is especially telling:

> For if some architect conceives a building in an orderly fashion, then although such a building never existed, and even never will exist, still the thought of it is true, and the thought is the same, whether the building exists or not. (*TdIE* § 69; C, p. 31)

After this, partly relying on the traditional doctrine (uncontroversial still in the seventeenth century) according to which for each true essence there is a definition that fully captures that essence,[8] Spinoza makes explicit the essentialism involved in his epistemology: '[T]he best conclusion will have to be drawn from some particular affirmative essence, *or*, from a true and legitimate definition' (§ 93; C, p. 39). Thus, understandably, 'the right way of discovery is to form thoughts from some given definition. This will proceed the more successfully and easily, the better we have defined a thing' (§ 94; C, p. 39).

Obviously, Spinoza's position is that if we succeed in acquiring the true definition of an essence, we are in the desirable epistemic position he had earlier described as having a true idea from which 'objective [i.e. representational]

[7] Because of the pride of place given to essences by Spinoza, I would classify him, in terms promoted by Jorge Secada, as a proponent of 'essentialism', 'the view that knowledge of the essence of a substance is prior to knowledge of its existence' (Secada 2000, p. 1).

[8] Cf. especially 1p8s2, 1p16. Christia Mercer (2001, p. 227) characterizes a widely accepted seventeenth-century view as follows: 'First, an essence is what is given in the definition of the thing and what can be grasped by the intellect; second, it constitutes the nature of an individual and that from which its properties flow.'

effects proceed in the soul according to the formal nature of its object' (§ 85; C, p. 37). Our mind can reproduce 'objectively [i.e. representationally] the formal character of nature' (§ 91; C, p. 38). This kind of orderly conceiving of things can in no way lead us astray, for it only contains mental affirmations that track or accurately represent what they are ideas of, the being of essences. Interestingly, there seems to be a strong element of *compulsion* involved in all this; to borrow a phrase from the *Short Treatise*, 'it is the thing itself that affirms or denies something of itself in us' (*KV* II.16; C, p. 124; G I, p. 83). Obviously, when given a true idea, our mind has little choice over the fact that it engages in a certain kind of veridical process.

Now striking as the just outlined position may appear, it does not come out of thin air: its immediate background is Descartes's doctrine of true and immutable natures. The famous passage of the Fifth Meditation runs as follows:

> But I think the most important consideration at this point is that I find within me countless ideas of things which even though they may not exist anywhere outside me still cannot be called nothing; for although in a sense they can be thought of at will, they are not my invention but have their own true and immutable natures. When, for example, I imagine a triangle, even if perhaps no such figure exists, or has ever existed, anywhere outside my thought, *there is still a determinate nature, or essence, or form of the triangle which is immutable and eternal, and not invented by me or dependent on my mind.* This is clear from the fact that various properties can be demonstrated of the triangle, for example that its three angles equal two right angles, that its greatest side subtends its greatest angle, and the like; and since these properties are ones which I now clearly recognize whether I want to or not, even if I never thought of them at all when I previously imagined the triangle, it follows that they cannot have been invented by me. (CSM II, pp. 44–5, emphasis added)

At least three points concerning the close connection between Descartes and Spinoza can be immediately established. First, the essences (with their being) Spinoza speaks about are eternal and – just as they are for Descartes – *immutable*. We have already encountered the eternity claim, and in the *Short Treatise* Spinoza makes clear the point concerning immutability.[9] Second, neither thinker leaves unclear the veridical nature of these essences: after all, they are *true* and immutable for Descartes of the *Meditations*, and the basis of true cognition for Spinoza of the *Treatise on the Emendation of the Intellect*. Third, both Descartes and Spinoza argue that the essences in question are mind-independent, not just products of our imagination. As we have seen, Spinoza simply asserts without any further argument, 'a thing

[9] For example: '[T]he essence of a thing is immutable' (*KV* II.15; C, p. 121; G I, p. 80).

that is conceived clearly and distinctly, *or* truly, outside the intellect is something different from the idea' (*CM* 1.2; C, p. 305; G 1, p. 238), whereas Descartes offers – as we will see in more detail below – an argument based on the element of compulsion involved in cognizing the essences: whether the meditator wants it or not, he recognizes that certain properties belong to the triangle, which, Descartes argues, shows that the geometrical figure with its essence 'cannot have been invented' by the meditator.[10]

I think we can gain deeper insight into what motivates Spinoza's thought by taking a look at two aspects of Descartes's doctrine of true and immutable natures, without getting too deep into the still on-going debate over the nature of that doctrine. I propose we key into these aspects by framing two sets of questions. First, why does Descartes feel the need to introduce the true and immutable natures? What purpose do they serve, what is their task? Second, what is the ontological status of the true and immutable natures? What kind of entities are they, how do they fit into the Cartesian metaphysics as a whole? The first set of questions is, I believe, convincingly answered by the account given by John Carriero; that the latter set of questions has given rise to a lively dispute does not prevent them from shedding light on the kind of influence on Spinoza that the Cartesian doctrine might have had.

In his *Between Two Worlds*,[11] Carriero situates the argument Descartes presents in the Fifth Meditation by stating it to involve 'a reorientation in the methodology of essence': real cognition is not, as it was for the Aristotelians, based on sense experience of actual substances providing the material from which knowledge of the essences of those substances is obtained through a process of abstraction; rather, 'Descartes thinks we are in cognitive possession of the nature of body before we experience any bodies, indeed, before we know whether any bodies exist to be experienced.' According to Carriero, the need for the true and immutable natures stems mainly from the fact that the Cartesian scheme requires mind-independent entities ('somethings') for there to be the necessary foothold for true cognition: '[U]nless there are true things, to which

[10] However, in a marginal note to the *Short Treatise* we find a passage reminiscent of Descartes's argument:

> [W]hether they [ideas of things other than God] exist or not, their essence is always necessary like the Idea of a triangle [. . .], so that even if I thought at first that I had feigned them, afterwards I would still be forced to say that they are and would be no less the same, even if neither I nor any other man had ever thought of them. (*KV* 1.1; C, p. 63; G 1, p. 17)

This, I take it, is an important reason for regarding the entities under consideration as non-fictitious. Moreover, the example involving a mountain and a valley Spinoza gives slightly earlier in the *Short Treatise* (1.1; C, p. 61; G 1, p. 15) appears already in the Fifth Meditation (CSM 11, p. 46).

[11] Carriero 2009, p. 293.

we have cognitive access, we won't be able to enter into a relation of "adequation" with them, and we won't be able [to] form true judgments.'[12] In other words, developing an anti-Aristotelian epistemology that does not take sense perception as its basis leads Descartes to posit natures as entities that exercise, to borrow Carriero's expression,[13] proper 'control' over our cognitive processes; 'clear and distinct perception correctly tracks essences'.[14] This sounds plausible given that Carriero's overall characterization of Descartes's approach appears accurate: if our cognition is not sense perception-based as claimed by the Aristotelians, without the fixed extramental points of reference provided by the true and immutable natures the use of our cognitive abilities would risk becoming a bizarrely arbitrary affair.

Interestingly, Descartes's argument for the mind-independence of the natures his new theory requires turns on a central element of the traditional philosophical landscape. As Carriero explains,[15] an essence forms the core of what it is to be a certain thing, and in virtue of its essence alone, the thing has certain necessary properties of which we can be *unaware* even while having a cognitive access to the essence itself (that is, it is possible to know that something has the essence definable as *a closed plane figure formed by three intersecting lines* without knowing that in virtue of that essence the thing in question necessarily has the property of having internal angles summing to two right angles).[16] Now Descartes claims that precisely this shows that an entity we are cognizing is a non-fictitious one: an arbitrarily composed thing (such as a chimera) cannot be the source of such unforeseeable properties.

We have thus, I take it, established the task of the immutable natures (they offer the crucial bridgehead for our cognition) and the argument for their mind-independence (based on them having unforeseeable necessary properties). What kind of entities are they, then? This is where the road of textual evidence ends and a lively interpretative debate commences. Given that the meditator says that a true and immutable nature is an eternal entity 'not invented by me or dependent on my mind' (CSM II, p. 45), it is no surprise that Descartes has been interpreted, most notably by Anthony Kenny, as championing an essentially Platonist position.[17] For our purposes, however, there is no need to take a stand on this debate – and it might

12 Carriero 2009, p. 311. It should be noted that Carriero (ibid., p. 283) also points out that were there no true and immutable natures, 'it is not clear that there would be anything for God to create'. Of course, the true and immutable natures also are the basis of the ontological argument for God's existence that Descartes presents in the Fifth Meditation.

13 See ibid., pp. 293–4, 299. 14 Ibid., p. 282. 15 Ibid., pp. 300–4.

16 For more on this kind of view of essences and properties, see below. Cf. also Wilson 1978, pp. 171–2.

17 Kenny (1970, pp. 692–3) writes: 'Descartes's theory [. . .] is one of nonexistent objects that *have* essences. [. . .] Descartes's philosophy of mathematics [. . .] is thoroughly Platonic: indeed he is the

well be the best move to refrain from taking a definite stand on it –[18] for here it suffices to make a more modest related point. Namely, whatever the final ontological status of the Cartesian true and immutable natures may be, there is no question that the doctrine may push the reader to think along rather Platonic lines. Kenny delineates a traditional view he calls 'Platonism about essences' which 'has similarities with that of Descartes'; according to it, there also are such things as 'real beings without existence'.[19] He explains:

> Even without existence, a real being differs from a fictional being, in that it *can* have existence and therefore it has a certain absolute reality before it exists. This absolute reality is called '*esse essentiae*': it reminds a modern reader of the status of Meinong's pure objects, beyond being and nonbeing. This belongs to it because of its relationship to an exemplar in the divine mind: just as God is the efficient cause of the existential being of things, so he is the exemplar cause of their essential being.[20]

This view may well be to a certain extent similar to that of Descartes; even more strikingly it resembles what Spinoza asserts in the early passages quoted above. The line of division between reality and fiction is drawn so that concrete existence is not required of real beings; rather, any true essence is endowed with an absolute (i.e. mind-independent) form of reality of its own called *esse essentiae*, being of essence. Of course, Spinoza would not accept the exemplarity thesis according to which essences are what they are because they somehow exemplify God's mind: as the above-quoted passage from the *Metaphysical Thoughts* 1.2 reveals and as we will see in more detail below, he views the relationship between God and the being of essences in a different way. Moreover, it is not altogether clear which things, for Descartes, have eternal essences; in a letter to Mersenne he mentions God, mind, body, and triangle (CSMK, p. 183).[21] Spinoza's answer is unequivocal: God and all the modes of his attributes.

To sum up the discussion thus far, we can discern two components in Spinoza's theory of essences, the ontological and the epistemological, and find a way to relate them to the thought of his predecessors. With regard to the ontological component, Spinoza's thought belongs to – and may in fact

founder of modern Platonism.' Cf. also Kenny 2009 (1958), ch. 7. For discussion and differing views, see Wilson 1978, ch. 5; Schmaltz 1991; Nolan 1997; Cunning 2003; Doney 2005; Abbruzzese 2007; Carriero 2009, ch. 5.

[18] This is what Carriero (2009, ch. 5) does. However, I think it is quite clear that the natures in question belong to *things*, not to our ideas of them; cf. Carriero 2009, p. 467. Moreover, I do not see why Descartes's dualism could not accommodate true and immutable natures, if they are seen to include essences 'of *certain kinds of extended beings* – for example, cubes, spheres, squares, circles, etc.' (Carriero 2009, p. 292, emphasis added).

[19] Kenny 1970, p. 696. [20] Ibid. [21] Cf. Secada 2000, p. 67.

be the high point of – an essentialist tradition that originates with Plato and functions as a shared background for the scholastics (and perhaps for Descartes, too, although this is a debated issue). On Spinoza's view, the being of essences constitutes the pre-eminent layer of reality that determines, eternally and immutably, the way things are. With regard to the epistemological component, the new Cartesian, anti-Aristotelian cast of mind dominates: true cognition of things proceeds through essences to which we can gain immediate cognitive access; sense perception of actually existing things is cognitively posterior.

Hence, on this fundamental level of his philosophical thinking Spinoza is, as it were, Descartes turned monist and, at least in an important sense, Platonist.

One absolutely focal aspect of Spinoza's theory of essences is, however, as yet undiscussed, and the rest of this section is devoted to it. We noted above that Descartes presents what might be called the argument *from unforeseeable properties* to back up his claim that true and immutable natures cannot be mind-dependent. This argument is based on an understanding of essences that is in an important sense completely traditional. Any true essence, whether of a natural agent, God, or a geometrical entity, is such that from it certain necessary properties follow, flow, or emanate (whatever the exact nature of this may be). However, there is something special in the passage of the Fifth Meditation that I want to emphasize: although it deals with a potentially unlimited range of different kinds of things with essences, a *geometrical figure* is taken as the representative of all things.[22] No doubt, here we encounter an important point of contact between the Cartesian ontology and the Galilean view of nature cited in the beginning of the introduction. Consider what Descartes writes in the subsequent Sixth Meditation:

> They [corporeal things] may not all exist in a way that exactly corresponds with my sensory grasp of them, for in many cases the grasp of the senses is very obscure and confused. But *at least they possess all the properties which I clearly and distinctly understand, that is, all those which, viewed in general terms, are comprised within the subject-matter of pure mathematics*. (CSM II, p. 55, emphasis added)

By rejecting substantial forms Descartes may have, as has been pointed out, rejected essences or substantial forms as causally robust entities required for

[22] For instance Margaret Wilson (1978, pp. 169–71) reads this passage from the point of view of Descartes's philosophy of mathematics; and relevant for it as the claims here no doubt are, it should nevertheless be kept in mind that the geometrical example is precisely that, an example – albeit of an especially noteworthy kind.

physical explanations;[23] but the basic *metaphysical* task they also were supposed to carry out, namely the task of individuating things,[24] had anything but vanished. I believe Spinoza was very much alive to the fact that in terms of philosophical explanation, there must be an ontological factor responsible for making the world and its inhabitants what they are; what is more, he models these crucial ontological features after the essences of geometrical entities unique in the intellectual grasp they can offer us of the immutable necessity with which particular (non-essential) properties accompany certain essential features. Descartes's way of thinking no doubt made the deepest impact on the young Spinoza, perhaps giving rise to something akin the following question: what if *each and every thing*, as also suggested by the Galilean view of the physical nature, is endowed with *essentially the same kind of internal structure* as geometrical entities are?[25] I think that, little by little, this idea grew stronger in Spinoza, finally resulting in a full-blown conviction: many subtleties of the Aristotelian-scholastic philosophy can and must be cast aside, for all things are intelligible to the very bottom just as things are intelligible in geometry. Thus the Cartesian spirit becomes radicalized as it were, its doctrine of material nature generalized and implanted into a framework of potentially unlimited number of basic attributes of the unique substance, the modifications of which are finite things.

On the present interpretation, we should appreciate that much in Spinoza's doctrine of being revolves around the thesis that an eternal essence makes a thing what it is; to that essence certain properties pertain, whereby becomes determined the thing's unique internal ontological make-up or structure – the form its being takes. Recall the argument in the Fifth Meditation: it indicates that there is not much if anything in this thesis that either Aristotelians or Cartesians would oppose. In its basics, the idea is quite accessible and attractive enough to motivate much of Western thought: each genuine thing has certain features constitutive of it, forming the core of the thing's being (its nature or essence), and in virtue of being an entity of a particular kind of nature the thing always has some further features that characterize it (its necessary properties). This line of thought seems to be a kind of common factor in Spinoza's philosophical background; it is only

[23] In her very instructive recent study *Descartes on Forms and Mechanisms* (2009), Helen Hattab argues for this and sheds light on the philosophical and historical factors contributing to Descartes's rejection of substantial forms. See also Kenny 1970, p. 698, and Ch. 3 below.

[24] As Hattab (2009, pp. 2–3) puts it: 'At the metaphysical level the substantial form accounts for the individuation of substances, and their identity over time.' See also Des Chene 1996, p. 54. For more on individuation, see Ch. 3 below (n. 68).

[25] See Carriero (2009, pp. 292–5) for Cartesian essences as real structures.

when one has to designate what kind of entities count for essences and properties, and what kind of relation holds between the two categories, that disputes begin. Spinoza simply lays emphasis on the universally acknowledged close and intelligible linkage between a geometrical entity's essence and its properties, taking this as the model of the inner ontological structure of any true thing.

Spinoza puts the geometry-inspired model to frequent use throughout his corpus, but there are a couple of particularly illuminating and philosophically important passages that should be taken up at this point. When he discusses in the *Treatise on the Emendation of the Intellect* the nature of definitions that describe or express essences,[26] Spinoza makes the following assertion:

> We require a concept, *or* definition, of the thing such that when it is considered alone, without any others conjoined, all the thing's properties can be deduced from it (as may be seen in this definition of the circle). For from it we clearly infer that all the lines drawn from the center to the circumference are equal. (*TdIE* § 96; C, p. 40)

Strikingly, '[t]hat this is a necessary requirement of a definition is so plain through itself to the attentive that it does not seem worth taking time to demonstrate it' (*TdIE* § 96; C, p. 40). This basic ontological structure pertains, for Spinoza, so evidently to all things that he appears to be on the verge of not mentioning it at all when he lists the requirements for the definition of an uncreated thing:

> Finally (though it is not very necessary to note this) it is required that all its properties be inferred from its definition. (*TdIE* § 97; C, p. 40)

And then later, having proven in the opening part of the *Ethics* that God is the only substance there is, Spinoza puts forth a proposition as crucial as it is intriguing, and one whose demonstration turns on the fact that also God is endowed with the ontological architecture discussed here:

> From the necessity of the divine nature there must follow infinitely many things in infinitely many modes (i.e., everything which can fall under an infinite intellect). (1p16)
>
> This Proposition must be plain to anyone, provided he attends to the fact that *the intellect infers from the given definition of any thing a number of properties that really do follow necessarily from it (i.e., from the very essence of the thing)*; and that it infers more properties the more the definition of the thing expresses reality, i.e., the more reality the essence of the defined thing involves. But since the divine nature has absolutely infinite attributes (by d6), each of which also expresses an essence

[26] Spinoza's theory of definitions will be discussed in more detail below; see Ch. 4 below.

infinite in its own kind, from its necessity there must follow infinitely many things in infinite modes (i.e., everything which can fall under an infinite intellect), q.e.d. (1p16d, emphasis added)

We will return to this proposition many times over; for present purposes it suffices that precisely what is here left implicit should now be evident.

Clearly, as already noted, Spinoza's theory of being derives much of its inspiration from the new Galileo–Cartesian picture of material nature. However, Spinoza consistently stresses – this forms a major part of his radicalism – that the geometry-derived model applies, precisely the same way it does to corporeal things, to *all* things, thus to everything mental as well. If there is any doubt of this, consider the *Treatise on the Emendation of the Intellect*: it ends with an attempt to define the nature of the mind which is based on the assumption that the intellect has the kind of structure we have been discussing:

> Since the chief part of our Method is to understand as well as possible the powers of the intellect, and its nature, we are necessarily forced, by what I have taught in this second part of the Method, to deduce these from the very definition of thought and intellect. (*TdIE* § 106; C, p. 43)

> But so far we have had no rules for discovering definitions. And because we cannot give them unless the nature, *or* definition, of the intellect, and its power are known, it follows that either the definition of the intellect must be clear through itself, or else we can understand nothing. It is not, however, absolutely clear through itself; but because its properties (like all the things we have from intellect) cannot be perceived clearly and distinctly unless their nature is known, *if we attend to the properties of the intellect that we understand clearly and distinctly, its definition will become known through itself. We shall, therefore, enumerate the properties of the intellect here.* (*TdIE* § 107; C, p. 43, latter emphasis added)

After this, Spinoza lists eight properties, such as '[i]t forms positive ideas before negative ones' (*TdIE* § 108; C, p. 44), and then the work abruptly breaks off with the words,

> [i]t [the essence of thought] is rather to be sought from the positive properties just surveyed, i.e., we must now establish something common from which these properties necessarily follow, *or* such that when it is given, they are necessarily given, and when it is taken away, they are taken away (*TdIE* § 110; C, pp. 44–5),

never arriving at the sought-after definition or essence of the intellect. The whole attempt, however, clearly is about uncovering the essence/property structure with a specific kind of procedure that exemplifies what Piet

Steenbakkers convincingly identifies as Spinoza's very understanding of method: a combination of both analysis (moving from effects to causes to arrive at simple propositions) and synthesis (moving from causes to effects).[27] Recall that the properties of a thing 'follow' or 'flow' from its essence, which obviously means that they are in some sense causally responsible for those properties.[28] Spinoza acknowledges that 'true knowledge proceeds from cause to effect' (*TdIE* § 85; C, p. 37), and he obviously is trying (1) to list the necessary properties of the intellect, (2) by inspecting the properties, to arrive at the true definition of the intellect, and (3) to derive the properties from the acquired definition.[29] However, he does not get past the first stage.

Spinoza's treatment of the entities of the mental realm betray one of his greatest discontentments with Descartes's philosophy: the idea of humans as beings endowed with a mind, a special kind of substance that does not obey the same kind of laws as all natural (material) things do:

> Most of those who have written about the Affects, and men's way of living, seem to treat, not of natural things, which follow the common laws of nature, but of things which are outside nature. Indeed *they seem to conceive man in nature as a dominion within a dominion*. For they believe that man disturbs, rather than follows, the order of nature, that he has absolute power over his actions, and that he is determined only by himself. [. . .] It is true that there have been some very distinguished men (to whose work and diligence we confess that we owe much), who have written many admirable things about the right way of living, and given men advice full of prudence. But no one, to my knowledge, has determined the nature and powers of the Affects, nor what, on the other hand, the Mind can do to moderate them. I know, of course, that the celebrated *Descartes, although he too believed that the Mind has absolute power over its own actions, nevertheless sought to explain human Affects through their first causes, and at the same time to show the way by which the Mind can have absolute dominion over its Affects. But in my opinion, he showed nothing but the cleverness of his understanding*, as I shall show in the proper place. (3pr; C, pp. 491–2; G II, pp. 137–8, emphases added)

As Spinoza notes, Descartes discusses our mental, especially emotional, life; but clearly, Spinoza does not find that discussion adequate,[30] and he gives

27 Steenbakkers 2009b, pp. 48–9. Compare the way in which Spinoza combines analysis and synthesis to the Aristotelian *regressus* proof in which 'one begins with observed effects, and reasons to their possible cause(s). Then, after a mental examination of the cause(s), one deduces the observed effect from the proper cause' (Hattab 2009, p. 113; cf. also p. 118).

28 More on this in the next chapter.

29 Cf. Joachim (1993 [1940], p. 200): '[W]e must always begin by forming a "good" definition, and always proceed by "deductive" development of its implications.'

30 The criticism seems to be directed to a great extent against the *Passions of the Soul* §§ 45–50 (CSM I, pp. 345–8).

the impression that he will be doing something different and new when he declares that our affects (i.e. emotions) acknowledge

> certain causes, through which they are understood, and have certain properties, as worthy of our knowledge as the properties of any other thing [. . .]. Therefore, I shall treat the nature and powers of the Affects, and the power of the Mind over them, by the same Method by which, in the preceding parts, I treated God and the Mind, and I shall consider human actions and appetites just as if it were a Question of lines, planes, and bodies. (3pr; C, p. 492; G II, p. 138)

That he thinks even such putatively hard to define things as emotions should be studied in this way is, I think, a clear indication, as well as an important facet, of Spinoza's geometry-inspired ontology. Spinoza's approach is understandable, given the presupposition that such mental entities as emotions of love and hate are constituents of a definite internal structure, the basic character of which is common to all things.[31] Apparently, then, Descartes goes wrong in disregarding that internal structure and in claiming that there could be such a thing as the 'absolute dominion' of the mind – endowed with free will – over the passive emotions. Because those emotions are effects of real causes with definite patterns of operation, complete control over our emotions would, in fact, require nothing less than a sudden and inexplicable annihilation of those causes – an incomprehensible rupture in the very make-up of things.[32]

As the seminal proposition 16 of the first part of the *Ethics* already reveals, the thesis concerning the internal architecture of things remains emblematic for Spinoza's mature thought too, forming an important part of his cast of mind. But the discussion thus far has focused on early material, and Spinoza's *magnum opus* presents his essentialism in an elaborated form. At the same time, Spinoza offers us a better fleshed-out view of the two layers of reality, the eternal and the temporal. So it is to the *Ethics* that we must now turn.

ETERNITY AND TEMPORALITY IN THE *ETHICS*

Judging from what he says about the being of essences, it seems evident that Spinoza's system accommodates a governing assumption inherited from the ancient Greeks: the world is not a chaotic flux but an orderly whole, and this

[31] For more on Spinoza's geometrical treatment of emotions, see Ch. 6.

[32] It should be noted that since according to Descartes (CSMK, p. 183) we do have *an innate idea of the mind* (i.e. of the mind's true and immutable nature), presumably with an internal structure reflecting the triangle example of the Fifth Meditation, Spinoza may be viewed as carrying out in full the Cartesian project (albeit in a framework denying substance pluralism).

could not be without *something* offering the crucial ontological bedrock, determining the nature of reality; that something is provided by the *essences*. New are the ideas that those essences are structured as they are in geometry, and that there is only one substance the modifications of which all the other things, with their essences, are. In fact, we have already seen how these two Spinozistic ideas are combined in a seminal proposition of his masterpiece, 1p16, according to which it is a fundamental fact about finite things that they are modifications of God from whose nature they follow as properties – 'by the same necessity and in the same way' as in geometry (1p17s). All this pertains to the most general metaphysical level, and is put in conspicuously atemporal terms. However, there are not only eternal but temporal things as well, and Spinoza acknowledges this, never suggesting that temporal everyday existence is somehow unreal. Accordingly, a distinction I will discuss in what follows, between eternal and temporal existence, underpins much of the *Ethics*.[33] Moreover, it appears that Spinoza's way of drawing that distinction, with novel terminology, harks back to the early distinction between being of essence and being of existence.

The *Ethics* is famous for the dictum that the proper way of considering things is *under a species* or *aspect of eternity* (*sub specie aeternitatis*), which is to consider them 'insofar as they are conceived through God's essence, as real beings, *or* insofar as through God's essence they involve existence' (5p30d). Respectively, considering things temporally, or insofar as they are 'under duration', is to 'conceive them to exist in relation to a certain time and place' (5p29s). It should be noted that Spinoza is here talking about considering or seeing things in a certain way, under a certain form or aspect. How could this difficult doctrine be further explicated? Thomas Lennon has recently called this the *sub specie*, i.e. 'under the aspect of', model derived from geometry and illustrates it by the following example. It is possible to consider one and the same triangle as an isosceles triangle (i.e. *sub specie* having two equal sides) or as a triangle whose angles at the base are equal (i.e. *sub specie* having equal base angles): '[C]onstructing an isosceles triangle just is to construct a triangle whose base angles are equal, and to think about the one is to think about the other, albeit in a different way, for they are the same thing.'[34] So, just as it is possible to consider geometrical objects under different aspects, it is possible, according to Spinoza, to consider any finite thing as eternal or as durational; and as both seeings-as are evidently veridical, things really are both. If this sounds surprising, it is obviously

[33] Cf. Wilson 1996, p. 97. [34] Lennon 2005, p. 13.

only because we are most often trapped in our temporal viewpoint, which keeps us from seeing the eternal in us.

The important thing for our purposes is that in the *Ethics*, the focal type of entity discoverable under the aspect of eternity is what Spinoza calls the formal essence (*essentia formalis*) of a thing. Understanding the nature and function of formal essences is, I believe, the key to understanding Spinoza's view of the eternal realm. In fact, this is not the first time we encounter these atemporal entities: an above-quoted passage from the *Metaphysical Thoughts* maintained that

> the formal essence neither is by itself nor has been created, for both these presuppose that the thing actually exists. Rather it depends on the divine essence alone, in which all things are contained. So in this sense we agree with those who say that the essences of things are eternal. (*CM* 1.2; C, p. 305; G 1, pp. 238–9)

Obviously, then, when early Spinoza is talking about the being of essences, he is talking about the being of *formal* essences[35] – which is helpful, given that the formal essences of the *Ethics* have proven to be something of a puzzle. One thing is beyond doubt, however: whenever Spinoza explicates the nature of formal essences, he stresses the intrinsic connection these eternal entities have to God's essence or attributes (which constitute that essence, as 1d4 states). This is what he does in the quoted early passage, and the most revealing piece of textual evidence in Spinoza's masterpiece follows suit:

> The ideas of singular things, *or* of modes, that do not exist must be comprehended in God's infinite idea in the same way as the formal essences of the singular things, *or* modes, are contained in God's attributes. (2p8)

Also the following should be considered:

> In addition to these two kinds of knowledge, there is [. . .] another, third kind, which we shall call intuitive knowledge. And this kind of knowing proceeds from an adequate idea of the formal essence of certain attributes of God to the adequate knowledge of the [NS: formal][36] essence of things. (2p40s2)

The former passage makes it clear that formal essences are contained in God's attributes, while the latter suggests that intuitive knowledge has them as

[35] Also Don Garrett (2009, p. 287) notes that formal essences resemble the Cartesian true and immutable natures (as well as Leibnizian essences).

[36] It should be noted that the additional 'formal' here does not carry much weight: Fokke Akkerman has shown that the additions found in *De Nagelate Schriften* are not Spinoza's but glosses made by his circle of friends and incorporated into the Dutch translation by Glazemaker; on this, see Steenbakkers 2009a, pp. 38–40.

objects. It is possible to proceed from the idea of an attribute to the ideas of essences of finite things presumably because those things, with their essences, follow from their attributes.[37] All this, especially the passage concerning the third kind of knowledge, is notoriously difficult. However, based on the textual evidence presented in this section I would suggest that what is revealed when things are conceived under a species of eternity, in the highest attainable way, are precisely formal essences – the eternal and immutable essences of things.[38] Now recall that these essences are endowed with their own kind of being and a specific resulting ontological structure. Given this, certain things can be said to follow from every essence,[39] which makes it possible to derive from the formal essence the complete layout, so to speak, of the thing in itself, as it would exist in complete isolation. This agrees with Charlie Huenemann's contention, 'formal essences [. . .] lack [. . .] extrinsic properties'; for instance a human being conceived as a formal essence 'is conceived in isolation from his particular surroundings, circumstances, ancestors, and descendants'.[40] An entity constituted by a formal essence is thus a thing whose being is autonomously determined by its own essence alone. It seems that here we can see, once again, the profound impact geometry has on Spinoza's mindset: the formal essence appears to match a geometrical object's essence in virtue of which certain properties pertain to the object in question.

When Spinoza illustrates these ontological and epistemological tenets, he relies, in keeping with the geometrical cast of mind, on examples taken from mathematics. He regards intuitive knowledge as a direct grasp of things and their properties, comparable to attaining mathematical insight:

> Suppose there are three numbers, and the problem is to find a fourth which is to the third as the second is to the first. [. . .] Given the numbers 1, 2, and 3, no one fails to see that the fourth proportional number is 6 – and we see this much more clearly because we infer the fourth number from the ratio which, *in one glance*, we see the first to have to the second. (2p40s2, emphasis added)

Alas, as an elucidation this only goes so far. Evidently, the sort of insight Spinoza is trying to depict is further articulable only uneasily if at all, almost as if refusing to be put into words. Margaret Wilson suggests, correctly I think,

[37] For more on the way in which things follow from attributes, see Ch. 3 below. See also Wilson 1983, pp. 182–5.

[38] On the eternality of formal essences, see Donagan 1979 (1973), pp. 250, 255; Koistinen 1998, pp. 74–5; Garrett 2009.

[39] Cf.: '[I]t seems to be his [Spinoza's] view that the most fundamental knowledge of singular things is of their "inmost essence," from which their properties flow' (Wilson 1996, p. 117).

[40] Huenemann 1999, p. 235.

that 'the second kind of knowledge differs from the third both in requiring steps of reasoning, as distinct from direct mental vision, and in failing to arrive at the inmost essences of things';[41] to this I would only want to add that the essences in question are eternal formal essences. Moreover, they are contained in God's attributes. The crucial example of this is drawn from geometry:

> [T]he circle is of such a nature that the rectangles formed from the segments of all the straight lines intersecting in it are equal to one another. So in a circle there are contained infinitely many rectangles that are equal to one another. Nevertheless, none of them can be said to exist except insofar as the circle exists, nor also can the idea of any of these rectangles be said to exist except insofar as it is comprehended in the idea of the circle. (2p8s)[42]

Edwin Curley states this example in a more accessible way: 'If AC and FG are any two lines intersecting at a point B in a circle, then the rectangle with base AB and height BC is equal in area to that with base BG and height BF.'[43] What Spinoza seems to be claiming is that just as any circle can be said to contain infinitely many rectangles of clearly definable nature, each attribute contains all the formal essences of the finite things that fall under the attribute in question; as Huenemann helpfully explains, the containment of rectangles in the circle is geometrical in the sense that 'when X geometrically contains Y, it means that X has sufficient features for producing Y, in accordance with sanctioned means of construction'.[44] And just as, by following this method, an infinity of rectangles can be derived from the idea of a circle, infinitely many things, each one with its own formal essence, follow from any attribute. Also this illustration underscores the point that formal essences are to be conceived in the mould of geometrical objects' essences from each of which an array of properties necessarily follow.

[41] Wilson 1996, p. 118.

[42] A relevant earlier and non-geometrical illustration runs:

> Finally, if any Philosopher still doubts whether essence is distinguished from existence in created things, he need not labor greatly over definitions of essence and existence to remove that doubt. For if he will only go to some sculptor or woodcarver, they will show him how they conceive in a certain order a statue not yet existing, and after having made it, they will present the existing statue to him. (*CM* 1.2; C, p. 305; G 1, p. 239)

For an instructive discussion on this example, see Koistinen forthcoming$_a$.

[43] C, p. 452 (a note to 2p8s); see also my illustration of Curley's point in Figure 1.

[44] Huenemann 1999, p. 233.

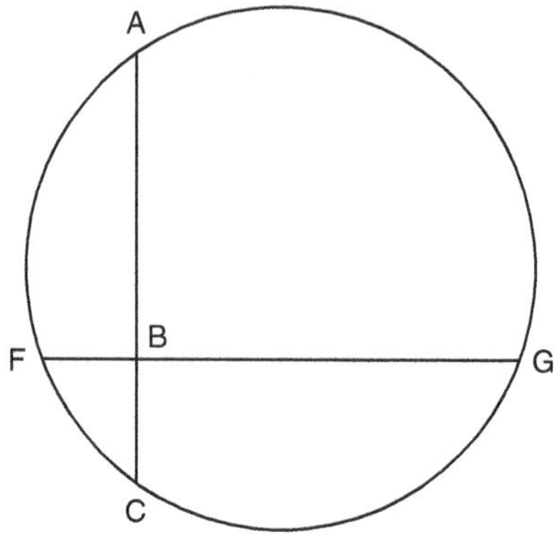

Figure 1

The geometrical example of the attribute/modification ontology is particularly striking when compared to the considerably more traditional relationship obtaining between Cartesian essences and God as their efficient cause – be the exact nature of that relationship ultimately what it may. Spinoza's conception may duly be called immanentist, for it recognizes no ontological gulf between God and creatures; rather, the latter are modifications or *states of attributes* constituting *the divine nature* itself. On this radical position Spinoza never wavers; recall that as early as in the *Metaphysical Thoughts* he says that an essence depends 'on the divine essence alone, in which all things are contained' and that 'the essences of nonexistent modes are comprehended in their substances, and their *being of essence* is in their substances' (*CM* I.2; C, p. 305; G I, p. 239). In fact, this idea lies at the core of Spinoza's naturalism: we are not dealing with transcendent God but with Nature itself, the modifications of whose essence are finite things.

Despite the undeniable distance between Descartes's and Spinoza's views on the relationship obtaining between God and finite things, a case can be made that here also Spinoza's thought displays strong Cartesian overtones. Namely, the geometrized understanding of the divine attribute/modification relationship may well be a *generalization* and an *elaboration* of a theory of the material universe Descartes presents in *The World*, provided that Helen Hattab's following recent interpretation of that work is accurate:

> Physically speaking, matter is not pure extension, but delimited extension, in the sense that God has attributed to it the basic divisions, proportions, motions, and relations that give rise to the particular shapes and motions we observe.[45]

[45] Hattab 2009, p. 148. See also pp. 143, 147, 152.

On this reading, Descartes redefines 'the object of physics as reified delimited quantity'.[46] Now for Spinoza, limitedness is the mark of finitude (1d2), and it can be said that particular bodies arise from the infinite extension being limited in specific ways;[47] a thesis that bears a clear resemblance to Hattab's Descartes. Interestingly, she claims that for Descartes, '[t]here is nothing about matter's extendedness that dictates that its particles take on the particular shapes, sizes, and motions that they do'.[48] But here Spinoza surely would object: particular modes of extension are what they are precisely because the attribute of extension 'dictates' them to be so – at least in this sense the Spinozist theory is an elaboration of its Cartesian predecessor. Further, *each and every attribute, also that of thought,* contains and 'dictates' all of its modifications – this is the generalization Spinoza makes. Of course, the circle example of 2p8s appears to apply best to the case of extension, but it is nevertheless supposed to reveal *sub specie aeternitatis* the basic nature of the relationship holding between *any* kind of divine attribute and the formal essences of its finite modifications.

The aforesaid goes also some way towards explicating Spinoza's strictly *a priori* – in both the old and the current sense of the term – way of doing philosophy. After all, the things we are dealing with are not only, to borrow Descartes's words, 'objects of the intellect alone' (CSM II, p. 37)[49] – which, it should be stressed, does not detract from their reality – but also derivable from God's essence.[50] More to the point, acknowledging the latter fact is the necessary first step of all true philosophy. In the *Ethics*, Spinoza asserts that failing to see this has kept philosophers from seeing

> the [proper] order of Philosophizing. For they believed that the divine nature, which they should have contemplated before all else (because it is prior both in knowledge and in nature) is last in the order of knowledge, and that the things that are called objects of the senses are prior to all. That is why, when they contemplated

[46] Ibid., p. 152. According to Hattab (2009, pp. 105–9, 141–3, 147, 150–1), the background of this idea is formed by the distinction between absolute quantity (unchanging *metaphysical* matter) and delimited quantity of geometry (intelligible *mathematical* matter) brought to the fore by such figures as Joseph Blancanus (1566–1624), a scholastic mathematician of Descartes's time.

[47] As Alan Nelson (2005, p. 7) succinctly puts it: 'The perfect wholeness and simplicity of the idea of the infinite must have limitations imposed upon it in thought to arrive at accurate conceptions of finite things. Spinoza expresses this by saying that finite things "follow" from the infinite and must be "conceived through" it.'

[48] Hattab 2009, p. 148.

[49] As Spinoza puts it in a letter to Meyer, 'there are many things that can in no way be apprehended by the imagination but only by the intellect, such as Substance, Eternity, and other things' (*Ep*12; S, p. 789; G IV, p. 57).

[50] Cf. the discussion above on the procedure of synthesis as it is applied in the *Treatise on the Emendation of the Intellect.*

natural things, they thought of nothing less than they did of the divine nature; and when afterwards they directed their minds to contemplating the divine nature, they could think of nothing less than of their first fictions, on which they had built the knowledge of natural things, because these could not assist knowledge of the divine nature. So it is no wonder that they have generally contradicted themselves. (2p10cs)

Again, infinite (God) comes before finite. This is the background against which should be read the following *prima facie* rather striking statements that Spinoza makes in his correspondence: 'To your question as to whether I have as clear an idea of God as of a triangle, I reply in the affirmative. But if you ask me whether I have as clear a mental image of God as of a triangle, I reply in the negative. We cannot imagine God, but we can apprehend him by the intellect' (*Ep*56; S, p. 905; G IV, p. 261). 'For I do not presume that I have found the best philosophy, but I know that what I understand is the true one. If you ask me how I know this, I reply that I know it in the same way that you know that the three angles of a triangle are equal to two right angles' (*Ep*76; S, p. 949; G IV, pp. 319–20). Of course, even if one were convinced that the divine nature is something as accessible to us as the ideas of extension and thought, and that everything finite is contained in and follows from the infinite attributes just as a triangle's properties can be said to be contained in and follow from its nature, it is another and far more radical thing to claim – what the quoted passages also at least suggest – that one grasps *the precise way in which* finite things follow from the divine nature.

Not only (formal) essences follow from the divine nature, but also the way in which things with those essences determine each other. Let us begin with the systematic reasons for this: first, there must be a fundamental explanation for the fact that finite entities enter into relationships of interdetermination (i.e. into interaction) with each other and for the specific nature of those determinations; second, the geometrical model implies this, for a geometrical object (say, a right-angled triangle) has several properties (for instance the property of fulfilling the Pythagorean theorem and the property of having internal angles summing to two right angles), and these properties surely are in some kind of relation of interdetermination. In terms of textual evidence, there is much in the latter half of the opening part of the *Ethics* that revolves around this: '[A]ll things have been determined from the necessity of the divine nature [. . .] to produce effects in a certain way' (1p29d); and Spinoza makes it clear that these determinations result from being determined by other finite modifications (1p28).[51] Moreover, in

[51] Cf. also: '[F]rom the given divine nature both the essence of things and their existence must necessarily be inferred' (1p25s). 'So will does not pertain to God's nature any more than do the

a letter to de Vries he affirms that both things *and the affections of things* are 'eternal truths' (*Ep*10; S, p. 783), which I take to mean that also the way in which finite things become affected (i.e. determined) by other finite things follows from God's nature.

I would suggest that the *formal being* Spinoza discusses mainly in 2p3–p7 involves more than just the formal essences of finite things, namely also all the particular states resulting from intermodificationary determination relations, that is, from the way in which formal essences determine each other. Spinoza calls the mediate infinite mode of extension 'the face of the whole universe [*facies totius Universi*]' (*Ep*64; S, p. 919; G IV, p. 278), and I would be willing to propose that that infinite mode is a system constituted by the formal essences and formal being of all finite extended modes;[52] it thus contains all the determinations pertaining to each and every extended modification. There is also the mediate infinite mode of thought, and thus I would suggest that it contains all the formal essences of ideas and their determinations.[53] But the most important thing is that no matter under which attribute finite things fall, the formal essences and formal being of all finite modes – and thus the whole universe – to its finest detail, is fixed 'from eternity and to eternity', to borrow the words located in 1p17s.

The moves Spinoza makes within his essentialist framework emerge, then, from a central requirement which justifies calling him a 'rationalist': the dislike of brute facts for which there would not be a proper explanation – 'cause *or* reason' (1d11d2) – ultimately derivable from nothing less basic than the nature of reality itself. On the present interpretation, this leads Spinoza to the view that only eternal formal essences (with their accompanying ways of determining each other) that follow from the divine attributes can be the basis of reality: it is only because of this atemporal layer of quasi-geometrically structured entities firmly rooted in the very essence of the whole of nature itself that things are kept from falling apart so that they form, as was commonly taken for granted still a century and half after Spinoza, an ordered whole. Nothing deserves to be called real if it is not included in this infinite blueprint of reality Spinoza of the 1660s calls 'the being of essences', of the 1670s 'things considered under the aspect of eternity'. It is the eternal in us that determines what we are.

other natural things, but is related to him in the same way as motion and rest, and all the other things which, as we have shown, follow from the necessity of the divine nature and are determined by it *to exist and produce an effect in a certain way*' (1p32c2, emphasis added).

52 For similar views, see Matheron 1988 (1969), pp. 33–4; Huenemann 1999, p. 237.

53 Here 5p40s is especially interesting; see also Donagan 1979 (1973), p. 255.

It is not difficult to understand, in its rough outlines, the meaning of *actual existence* in Spinoza's system: it equals concrete existence that occurs in time and place. We conceive this kind of existence *sub specie durationis*, and to a great degree through our senses. The task of giving a detailed interpretative articulation of the nature of finite existence – and the related issue of the nature of the actual essence (*essentia actualis*) of finite things – is something I will take up later on; at this point it suffices that we discern the way in which actual being of concrete existents is connected to the formal essences and being of the eternal realm.

On the present interpretation, formal essences as atemporal entities hold a pre-eminent position in determining the character of reality in its entirety.[54] Given this, there must be something in formal essences and formal being that designates the way in which things unfold under duration. In other words, the eternal essences and the being determined by their interrelations must contain *features at least convertible into determinations of time and place*. In his early works, Spinoza does not offer us much by way of elucidation on this topic, but there is one particularly important passage that corroborates my view:

> [W]hen we say that God has decided that the triangle shall exist, we are saying nothing but that God has so arranged the order of nature and of causes that the triangle shall necessarily exist at such a time. So if we understood the order of causes as it has been established by God, *we should find that the triangle must really exist at such a time, with the same necessity as we now find, when we attend to its nature, that its three angles are equal to two right angles.* (*CM* 1.3; C, p. 309; G 1, p. 243, emphasis added)

This implies that there is no difference in the necessity with which things are produced *sub specie aeternitatis* and the necessity with which temporal determinations of actual existents become fixed – both are what they are because they follow, with the same kind of necessity we find in geometry, from God's nature.

These observations point towards an interpretation of formal essences that can shed additional light on this admittedly extremely difficult issue. Olli Koistinen has argued that the formal essences of finite things should be understood as *objects of truths* about finite modes. As such they are not spatiotemporally limited, which justifies considering them to be *infinite* modes.[55] Koistinen explains:

[54] Cf.: 'On the other hand, he [Spinoza] makes it equally clear [. . .] that the formal essence of a singular thing is *directly related* to the singular thing, and even provides a sense in which the singular thing itself can be said to have a kind of derivative being' (Garrett 2009, p. 287).

[55] Koistinen (1998, pp. 71–5) develops his interpretation of formal essences as infinite modes as an answer to the problem that arises due to the fact that Spinoza claims nothing finite to follow from God's absolutely infinite nature. For other views of formal essences as infinite modes, see Martin 2008; Garrett 2009.

> Let us suppose that Jones raised his hand in his bedroom 12.2.1995. [. . .] [B]ecause this hand raising has spatiotemporal limits it is a finite mode. But consider now the truth expressed by the sentence 'Jones raised his hand in his bedroom 12.2.1995.' This sentence is *about* the finite mode [. . .] [b]ut [. . .] what makes it true is not just the existence of 'Jones' raising his hand' but 'Jones' raising his hand in his bedroom at 12.2.1995.' [. . .] This entity is beyond the temporal and spatial order and is for that reason an infinite mode. But because all truths about finite modes must involve place and time specifications it follows that all truths about finite modes have as their objects infinite modes and are made true by infinite modes.[56]

On Koistinen's interpretation of Spinoza's necessitarianism, there must be omnitemporally existing entities that function as the bearers of truths concerning finite temporal existents. These entities Spinoza would then call formal essences, and a formal essence corresponds 'to the object of a truth about a finite mode. It is these formal essences or objects of truths about finite modes that follow from the eternal and infinite essence of God.'[57] To my mind this sounds importantly correct; I would only want to add that in addition to the formal essences – which, I think, designate things as they are in themselves[58] – there is formal being that consists of the determinations resulting from the relations the formal essences bear to each other. This is, however, a merely terminological issue; the important point concerns the determinative priority of eternity. Each and every truth about an actual thing has an eternal and unchanging object of truth that follows from God's nature – which also is, given that God is a necessary existent, the crux of Spinoza's necessitarianism.[59] Finite things are not necessary existents, but this does not mean that it would not be absolutely necessary that they exist when they do and the way they do.[60] Interestingly, this seems to be the Spinozistic – that is, the monist, immanentist, and necessitarianist – counterpart of a late scholastic position concerning the eternality of God's volitions and the temporal world, lucidly formulated by Dennis Des Chene: 'Of those propositions that are willed some are, from the point of view of a temporally ordered world, "executed" at a particular time and place. But that does not count against their having been eternally willed.'[61]

[56] Koistinen 1998, p. 73. [57] Ibid., p. 75. Cf. also Garrett 2009, p. 287.

[58] For more on this, see Ch. 3 below.

[59] As Koistinen (1998, p. 75) puts it, 'that they [the formal essences] follow from the eternal and infinite essence of the necessarily existing God is, in Spinoza's system, sufficient for there being only one possible system of finite modes'.

[60] Also in this I agree with Koistinen (ibid., p. 74) who, as I understand him, makes the same point, but in different terminology.

[61] Des Chene 1996, p. 319. Here especially Peter Fonseca (1528–99) is an important source for Des Chene.

The eternal and infinite systems of essences, together with the determinations specified by them, are thus converted into and correspond to the spatiodurational existence of the actual world.[62] As we will see in more detail below, actual finite existence has its peculiar character: in it, limitations are not mere determinations but *oppositions* and *agreements* that take place between *striving* entities. But here the main point concerning actual existence is that any actual thing's temporal path, forged from a specific set of affections, is decreed from, and thus conceivable under the aspect of, eternity. For humans that path may be one of innumerable tumults and conflicts – all of which, however, are based on the imperturbable rigour of the eternal and intelligible structure of being determined by God-or-Nature's essence itself.

[62] Cf.: '[W]e may also think of the actual essence of a singular thing as the *actualization* or instantiation of its existing formal essence, rendering the thing itself actual' (Garrett 2009, pp. 286–7).

CHAPTER 2

Causation and geometry

As we have seen, it is a central characteristic of Spinoza's rationalism that everything can – at least in principle – be explained; there are no brute facts. Moreover, everything has its *cause* through which it can be explained. Indeed, according to Spinoza, not only the existence of everything that exists but also the non-existence of everything that does not exist requires an explanation:

> For each thing there must be assigned a cause, *or* reason, as much for its existence as for its nonexistence. For example, if a triangle exists, there must be a reason *or* cause why it exists; but if it does not exist, there must also be a reason *or* cause which prevents it from existing, or which takes its existence away. (1p11d2)

The essentialist, rationalist, necessitarian, and monist system explicated above is designed to offer us a proper metaphysical picture of why things exist, and of the way they do. And since that system's specifically shaped entities – regardless whether we consider them *sub specie aeternitatis* or *durationis* – are real, what follows from their essences must be real effects. The ordered whole of real things is an infinite causal network. Reasons for thinking in this way will become clearer below, but on a general level it can be observed that this transition into causality evinces Spinoza's silent but persistent sensitivity to a focal intuition of the Aristotelian tradition: that real entities are causally efficacious, or powerful, ones. This means that the ontological structure of things is played out in the causal register, and understanding the nature of Spinoza's dynamic system must thus proceed through an accurate analysis of the view of causality it involves.

The axioms concerning causes and effects are located at the beginning of the *Ethics*:

> From a given determinate cause the effect follows necessarily; and conversely, if there is no determinate cause, it is impossible for an effect to follow. (1a3)

The knowledge of an effect depends on, and involves, the knowledge of its cause. (1a4)

So, causation involves necessity, nothing is outside of it, and effects are known through their causes. In addition to these highly abstract contentions,[1] the latter half of the first part of the *Ethics* deserves to be taken up, for it discusses first and foremost God's causality. In fact, a proposition we are already familiar with, 1p16, signals the beginning of the discussion. It would be difficult to overestimate the importance of that proposition, for in it and its corollaries Spinoza designates what his basic ontological tenets amount to when put in causal terms. Recall that the proposition '[f]rom the necessity of the divine nature there must follow infinitely many things in infinitely many modes (i.e., everything which can fall under an infinite intellect)' turns, to a great extent, on the contention that 'the intellect infers from the given definition of any thing a number of properties that really do follow necessarily from it (i.e., from the very essence of the thing)', which, in turn, is a compact formulation of Spinoza's essentialist understanding of the nature and structure of being. And then, with no forewarning, comes a statement concerning causality: 'From this it follows that God is the efficient cause of all things which can fall under an infinite intellect' (1p16c1).

How should all this be explicated? One useful way to study Spinoza's understanding of causality is to analyse it in relation to the Aristotelian-scholastic tradition, for centuries dominant enough to form the virtually ubiquitous backdrop of early modern philosophical thought. As is well known, that tradition relies on Aristotle's distinction of four basic kinds of cause: the material, the formal, the efficient, and the final cause. I will argue that comparing Spinoza's views to those of some prominent scholastics reveals that the basic model of causation cannot be, for Spinoza, one adopted from the mechanical sciences; what emerges from his ontological commitments is a model of causation heavily inspired by the idea of formal causality pertaining to geometrical objects.

THE FINAL, THE MATERIAL, AND THE EFFICIENT CAUSE

Spinoza uncompromisingly rejects any doctrine that assigns final causes to God: 'Nature has no end set before it, and [. . .] all final causes are nothing but human fictions' (1app; C, p. 442; G II, p. 80). There has been a lively discussion about teleology in Spinoza's thought, but even the most enthusiastic proponents of the teleological interpretation do not maintain that Spinoza would have accepted divine teleology.[2] Since I here concentrate mainly on Spinoza's

[1] As Wilson (1991, p. 133) notes, 'Spinoza says very little to elucidate directly the concept or concepts of causality he relies on.'

[2] See especially Garrett 1999; Lin 2006.

view on God's, or the substance's, causation, I will not pursue this issue further but simply hold Spinoza to be an anti-teleologist concerning God.[3]

The first thing to note about the material cause is that even though Spinoza talks about matter, he does not, to my knowledge, designate it as a cause. However, Huenemann has recently suggested that the Spinozistic extended substance could be regarded as a kind of *prima materia*.[4] This approach may, as Huenemann contends, help to fit Spinoza 'into some of the history of thought about prime matter'.[5] This is primarily so, I think, because the extended substance can be seen as a continuous field that is modified according to individual essences, resulting in actual physical things.[6] Still, the idea of prime matter is deeply embedded in the hylomorphic doctrine, according to which natural things consist of matter and form, and this is quite foreign to Spinoza's framework. Most importantly, no Spinozistic attribute can match prime matter as it was commonly understood, i.e. as devoid of all forms, because attributes such as thought and extension already specify certain fundamental manners of being. Moreover, if extension is interpreted as a kind of prime matter, on the basis of parallelism (2p7) the same applies to all attributes, making thought, too, a kind of prime 'matter', which sounds problematic. So, due to the lacking textual evidence and the difficulties this line of interpretation encounters, I would not be prepared to endorse the idea that the material cause would have found its way to Spinoza's philosophy; it seems that the most we can say is that concerning physical reality, the idea of a spatial field has in Spinoza's system a position somewhat reminiscent of the one held by the doctrine of prime matter in Aristotelianism.

What about efficient causation, then? According to the Peripatetic tradition, the efficient or moving cause is the agent that draws out the form from potency to act.[7] It is important to keep in mind that in medieval philosophy also the efficient cause is closely linked up with the final cause: as Thomas Aquinas explains in *The Principles of Nature* (henceforth *PN*), different kinds of causes are intertwined so that the final cause is the cause of all causality and all other causes, and this implies that the efficient cause, in launching the process of actualization, is always directed or inclined

[3] For a detailed discussion on Spinoza and teleology, see Ch. 5 below.

[4] Huenemann 2004. Des Chene (1996, p. 81) characterizes prime matter as follows: 'Prime matter is the stuff, whatever it may be, which, when joined with substantial form, yields an individual substance.' It is the material cause of things by being a component in complete substances.

[5] Huenemann 2004, p. 32.

[6] Here I have in mind Jonathan Bennett's (1984, pp. 81–110) 'field metaphysical' interpretation of the extended substance; see Ch. 6 below.

[7] See especially *The Principles of Nature* III.18; *SW*, p. 14.

towards an end (*PN* III.18–19, IV.24; *SW*, pp. 14–15, 19). Now, the efficient cause, of course, appears in many central early modern texts, Spinoza's included, but by contrast to the Aristotelians, such anti-teleologically inclined thinkers as Hobbes and Spinoza see efficient causation as 'blind' in the sense that efficient causes are never end-guided.[8] The upheavals that took place in seventeenth-century natural science had undeniably a profound effect on that period's philosophy, and obviously on grounds of this and Spinoza's tendency to separate efficiency from finality, he is often taken to view all causation along the lines of *mechanistic* efficient causation. For instance Bennett asserts, 'Spinoza argues that nothing has a final cause because everything has an efficient cause.'[9] He justifies this by citing the appendix of the first part of the *Ethics* where Spinoza denies final causes by invoking, among other things, 1p16 and by reminding us that everything in nature happens by a certain eternal necessity. As the following quote evinces, Bennett has a rather mechanistic way of looking at Spinoza's conception of efficient causality:

> The phrase about 'a certain eternal necessity of Nature' is a reference to Spinoza's efficient-cause determinism. He is implying that something which is caused mechanistically, i.e., by a 'push' from behind, cannot properly be explained also in terms of goals or purposes or desires, i.e., in terms of a 'pull' towards a resultant state of affairs.[10]

In general Bennett, like many others, seems to take for granted that the rejection of final causes directly entails that all causality is efficient causality, and since the paradigmatic case of efficient causality is usually considered to be the one that reigns in mechanics, a rather mechanistic picture of Spinoza's philosophy thereby emerges.

In a sense the mechanistically oriented interpretation of Spinoza's view on causation is thoroughly understandable, for not only was the science of mechanics prominent in the seventeenth century, but the importance of the efficient cause also emerges from Spinoza's texts; he states, for instance, that 'God is the efficient cause, not only of the existence of things, but also of their essence' (1p25), and that this efficacy takes place through an infinite chain of finite causes (1p28). Further, 'nothing belongs to the nature of anything except what follows from the necessity of the nature of the efficient cause' (4pr; C, p. 545; G II, p. 208). This indicates that any proper

[8] John Carriero (2005, pp. 121–2) expresses this point very instructively in his discussion on Spinoza and final causality. See also Carriero 1991, pp. 58–9. For a detailed examination of this, see Ch. 5 below.

[9] Bennett 1984, p. 215. [10] Ibid., p. 216.

interpretation of Spinoza's conception of causation must accommodate, and offer an explication of, the idea of efficient causality.

SUÁREZ ON THE FORMAL CAUSE AND EMANATION

Although there is no doubt that Spinoza includes efficient causation in his system, a quick look at the central 1p16 already reveals that it is hard to fit it into the mechanistic picture of efficient causality. In other words, judging from that proposition, the basic case of causation does not, for Spinoza, have to do simply with impacts through which motion is transferred from one body to another. Moreover and importantly, Spinoza's *order of presentation* – the fact that 1p16c1 says that God is the efficient cause *because* of 1p16 – speaks for this. So I suggest that we put the mechanism-associated way of thinking about efficient causation aside for a while and see how things look if we approach Spinoza's views by keeping in mind the scholastic conception of the formal cause and a related notion of emanation;[11] as we will see, a different picture emerges, and one that squares considerably better with the ontological considerations of the previous chapter. Thus, to obtain a better grasp of 1p16 and to see Spinoza's ideas concerning causation in their proper context, my next aim is to offer a brief explication of a relevant scholastic understanding of the formal cause and emanation.[12]

11 That the notion of emanation is relevant in interpreting Spinoza is, as such, by no means a novel observation: already Harry Wolfson (1961 [1934] 1, pp. 372–5, 391–2) lists what he sees as analogies between Spinoza's thought and certain aspects of the traditional doctrines of emanation. However, Wolfson's account of God's productive causality is hardly satisfactory; most notably, I do not think Spinoza is 'interposing infinite modes between God and finite modes' just as the emanationists interpose 'immaterial Intelligences between God and matter' (Wolfson 1961 [1934] 1, p. 391), for this kind of hierarchism, characteristic of the Neoplatonic doctrine of hypostases, is quite alien to Spinoza's system. Martial Gueroult (1968, pp. 246–52) maps Spinoza's relation to the traditional notion, and according to Carriero (1991), Spinoza operates within an Avicennan emanative framework. However, although important connections can be found between Avicenna and Spinoza, the idea of causation as emanation is not exclusively an Avicennan one. Moreover, Carriero seems to have a rather mechanistic interpretation of Spinoza's view of efficient causation. According to him, Spinoza applies 'the lessons of the new physics to human beings and the deity. [. . .] [T]he new science affords Spinoza powerful reasons for divorcing final causality from efficient causality and confining the former to an epiphenomenal status' (Carriero 1991, p. 59); in a later paper Carriero states that Spinoza 'presents a theory according to which the sort of causality the new scientists find in the corporeal order is found throughout all of nature' (Carriero 2005, p. 121). However, I think that precisely because Spinoza's conception of causation owes much to the idea of emanation its relation to mechanistic causation should be reconsidered.

12 There are, to my knowledge, two noteworthy readings that have regarded formal-emanative causality as focal for understanding Spinoza. According to Gueroult's classic study, Spinoza does not accept the traditional distinction between logico-mathematical emanation and active efficiency because he fuses the formal and the efficient cause together (Gueroult 1968, pp. 293–4, 297–9). Although Gueroult's work is very helpful, the statement that the formal-emanative and the efficient-active are assimilated

Of the scholastic philosophers, Francisco Suárez (1548–1617) offers us a helpful starting point. Suárez was, of course, the most prominent of the Renaissance Jesuit philosophers whose writings had a profound influence on post-Renaissance thought; indeed, medieval philosophy was passed, to an important degree, on to the modern world through his works. Descartes, who was trained in the scholastic tradition, probably had first-hand acquaintance with Suárez's *Disputationes*. In fact, it has been suggested that Suárez is the most important scholastic for Spinoza;[13] however, it should be noted that I am not attempting to establish a direct link between the two thinkers but merely taking Suárez as an instructive representative of a well-known line of thought in sixteenth-century philosophy – a line of thought in some important respects quite close to the claims contained in the *Ethics*. Recognizing this background helps us to see certain characteristics of Spinoza's thought that would otherwise easily go unnoticed.

The first thing to be kept in mind is that Suárez's conception of *natural* or *efficient emanation* seems to be very much unlike mechanistic efficient causation. Suárez's thought starts from the central scholastic distinction between necessary accidents (that is, necessary accidental properties, also known as proper accidents or *propria*) and non-necessary accidents; of these, non-necessary accidents can be taken away from a certain kind of substance (for instance, whiteness from a human being), while necessary accidents cannot (for instance, risibility from a human being).[14] According to Suárez,

(or, as Gueroult also says, that the latter is even in some sense reduced to the former) does not strike me as entirely accurate; a more explicative account of the relations between different kinds of causes in Spinoza's thought remains to be given. Moreover, Gueroult (p. 298) implies that the scholastics would not have seen emanation as true causal activity. But that this does not apply to all the prominent schoolmen – and the implications of this observation for interpreting Spinoza – will be presented below. Also Vincent Carraud (2002, esp. pp. 323–6) objects to interpreting Spinozistic causality as efficient causality; on his reading, proceeding through a discussion focusing on the expression *causa seu ratio*, it is formal causality that emerges as the basis of all causality. This is a good interpretative move, and congenial to the one I will favour below, albeit on different grounds. However, Carraud's reasoning leads him to analysing the formal conditions of the existence of things, whereas I would prefer a different approach to elucidate Spinoza's conception of causation. Moreover, I am unsure of how Carraud's view on the important relationship between formality and efficiency ('efficiency is just the external doublet of formality' [p. 324, translation mine]) should be understood. Although Wolfson mentions geometrical necessity while discussing emanation, he does not clarify its nature but merely proclaims, '[t]he term cause which Spinoza applies to God is [. . .] to be understood in the logical and geometrical sense' (Wolfson 1961 [1934] I, p. 373). So, although these scholars point in the right direction, I think a still more satisfactory account remains to be given.

13 Curley estimates, following Jacob Freudenthal's 'Spinoza und die Scholastik' of 1887, that 'of the better-known scholastics Suárez was probably the most important' (C, p. 223) for the medieval and late scholastic background of Spinoza's *Metaphysical Thoughts*. Referring to Curley, Lennon (2005, p. 27) writes that Suárez 'might have been his [Spinoza's] most important medieval source'.

14 See e.g. *PN* II.9; *SW*, p. 10.

substances produce their proper accidents through what he calls 'natural emanation':

> [T]he accidental properties, especially those that follow upon or are owed [to a substance] by reason of its form, are caused by the substance not only as a material cause and a final cause but also as an efficient cause through a natural resulting [. . .] [I]t is probable that *the substantial form has a certain power for having its proper accidents emanate from it.* Likewise, in this way one discerns more clearly *the natural connection between a* [substantial] *form and its properties.* (*MD* 18.3.4, emphases added)

To illustrate this, Suárez claims that water, even after having been heated, reduces itself to its 'pristine coldness' by 'the [substantial] form through a natural resulting' (*MD* 18.3.4), and this kind of principle of efficient causality is 'the inward substance itself' (*MD* 18.3.8). So, emanation is used to explain how things obtain their properties and why, in different situations, they act and react in certain characteristic ways: a thing's form and the properties resulting from it make the thing what it is.

If emanation has to do with formal causality, why, then, does Suárez call it *efficient*? As Des Chene explains, Suárez is, unlike the Thomists, decidedly of the opinion that emanation is genuine efficient action.[15] According to him, properties of natural things are distinct entities whose coming to being and persistence require a real action (*MD* 18.3.6–7); and this makes emanative causation a real action, even though it is not always, as Suárez acknowledges, counted as such (*MD* 18.3.6). In other words, for changes in the accidents of a substance (one, for example hotness, disappears, and another, for example coldness, appears) a cause, and hence real activity, is needed; and because, to use a traditional example, coldness emanates from the form of water even when water has been heated – bringing about a real change from hotness to coldness – the emanative production of properties must be, for Suárez, a true action.[16]

All this suggests that emanation is efficient because it refers to a true causal consequence or action in contrast to what Suárez calls a mere 'natural appropriateness' that is found, for example, between matter and form of celestial bodies – in them, matter and form are necessarily connected in a certain fashion, but there is no causality involved between form and matter (*MD* 18.3.6).[17] After all, Suárez's notion of the efficient cause seems to be rather broad, for he writes that the efficient cause causes 'by means of a

[15] Des Chene 1996, pp. 158–61. For discussion on Aquinas's position, see also Carriero 1991, pp. 69, 91–2.

[16] Since no thing can exist without its *propria*, it would probably be more accurate to speak here not about coldness but power to make colder (see Ch. 3, n. 7). However, there is no need for us to discuss this in detail here: it suffices to note that Spinoza's theory is subtle enough to give answers to this kind of question; see Ch. 6.

[17] For another reason for regarding emanation as real action, see Des Chene 1996, p. 160.

proper action that flows from it' (*MD* 17.1.6) and that 'to be an actual efficient cause is the same as being something that acts' (*MD* 18.10.5). Even though Suárez designates, in accordance with the tradition, the formal cause as an intrinsic, the efficient cause as an extrinsic cause (*MD* 17.1.6), he obviously thinks that the substantial form is distinct enough from certain accidents for it to be appropriate to think of emanation as one kind of efficient causation. So in its emanative action, a substantial form can be regarded not only as a formal but also as an efficient cause,[18] and Suárez certainly seems to think this to hold with regard to natural agents. This intertwinement does not, however, mean that the substantial form would be somehow unimportant or out of the picture: '*The substantial form is the principal principle by which the efficient cause acts*' (*MD* 18.2.3).

Recent scholarship indicates that the above claims were not contested in scholasticism, on the contrary. In an important paper tracing the role of the substantial form in the Aristotelian tradition, Robert Pasnau contends that '[f]or scholastic philosophers of all persuasions, the substantial form is the explanatory basis of the entire substance, serving as the internal cause of a thing's accidental properties and supplying the identity conditions for the whole substance and its parts.'[19] The following statement, in turn, offers a compact summary of the historical development of the notion: '[W]hereas in Aristotle a more metaphysical conception of form seems to predominate, by the end of the scholastic era the case for form rests entirely on its causal efficacy as the source of a substance's various intrinsic properties.'[20] Moreover, Pasnau discusses Suárez among those many thinkers who regard the substantial form as a robust causal agent or power, operating as a kind of internal efficient cause.[21] Clearly, then, what Suárez writes on emanation fits nicely into a widespread late scholastic line of thought.

The Suárezian view emerging from the preceding paragraphs can be summed up by saying that emanation is formal causation by which a thing produces its properties; in natural agents, this equals genuine causal efficacy. Martial Gueroult instructively observes that emanation was traditionally seen as immediate, with no mediating factors involved; it would thus in fact be contradictory to deny the emanative cause its effect so that the cause would exist but the effect not.[22] To take the traditional example

[18] Cf. Des Chene 1996, p. 332. [19] Pasnau 2004, p. 34. See also Nadler 1998, p. 515.
[20] Pasnau 2004, p. 41. [21] Ibid., pp. 37–8.
[22] Gueroult 1968, pp. 246, 269. As Gueroult notes, in this the emanative cause differs from the 'active' cause that produces its effect mediately (e.g. when a fire heats up a nearby table). For the corresponding distinction in Aquinas, see *SCG* II.30. See also Carriero 1991, p. 69.

Gueroult mentions, fire could not be fire without being hot,[23] so abolition of the hotness equals extinguishing the fire. Clearly, this kind of immediacy and necessity characterize what happens when a thing effects something in itself, in virtue of its own form alone. Moreover, the traditionally prevalent talk about *properties that follow from the essential principles* refers to this; here the key distinction is between a thing's essential principles or features (e.g. rationality of human beings) and the non-essential properties that necessarily follow from these essential principles (e.g. risibility of human beings).[24] In a nutshell, emanation means that from any thing's essence certain properties immediately and necessarily follow.

SPINOZA AND THE FORMAL CAUSE IN GEOMETRY

What does the preceding discussion imply with regard to Spinoza? At this point we should return to the focal 1p16, '[f]rom the necessity of the divine nature there must follow infinitely many things in infinitely many modes (i.e., everything which can fall under an infinite intellect)'. The demonstration turns on the tenet that 'the intellect infers from the given definition of any thing a number of properties that really do follow necessarily from it (i.e., from the very essence of the thing)'; and because 'the divine nature has absolutely infinite attributes (by d6), each of which also expresses an essence infinite in its own kind, from its necessity there must follow infinitely many things in infinite modes'. So there are things other than God, because God's essence is causally efficacious, bringing about all of God's properties. I think this notion of causality bears a striking resemblance to the above-outlined emanative action of the formal cause – moreover, a perfectly consistent resemblance given the fact that Spinoza's doctrine of being, as presented in Chapter 1, clearly has much in common with traditional essentialism. This impression is only strengthened by the final proposition of the first part of the *Ethics*: 'Nothing exists from whose nature some effect does not follow' (1p36). In other words, for Spinoza, things are *essential causers of properties*. Also, the *Theological-Political Treatise* displays these tendencies:

[S]ince the knowledge of the effect through its cause is *nothing other than the knowledge of a property of that cause*, the greater our knowledge of natural phenomena, the more

[23] Gueroult 1968, p. 246.

[24] See e.g. *SCG* II.30; Carriero 1991, pp. 51, 79; Garrett 1991, p. 201. As Carriero (1991, p. 73) argues, this kind of Aristotelian conception of essence differs from our contemporary way of construing a thing's essence out of its necessary properties. See also Lin 2004, pp. 26–7.

perfect is our knowledge of God's essence, which is the cause of all things. (*TTP* IV; S, p. 428; G III, p. 60, emphasis added)[25]

This way of putting things gestures towards the traditional view, such as the one endorsed by Suárez, according to which properties are caused by the substantial form from which they emanate.

A rather convincing case could be made, I believe, for there being a connection between formal causality and Spinoza's notion of formal essence – the latter seems, as we have seen, to be moulded after the geometrical objects and their essence/property structure, so perhaps calling the eternal essences precisely *formal* essences is a nod in the direction of the traditional notion of form, which also was considered a type of cause.[26] Be this as it may, at this point Spinoza's conspicuous eagerness to use geometrical objects in illustrating his ideas should be taken up. After having claimed that 'God acts from the laws of his nature alone, and is compelled by no one' (1p17) he remarks:

I think I have shown clearly enough (see p16) that from God's supreme power, *or* infinite nature, infinitely many things in infinitely many modes, i.e., all things, have necessarily flowed [*effluxisse*], or always follow, by the same necessity and in the same way as from the nature of a triangle it follows, from eternity and to eternity, that its three angles are equal to two right angles. (1p17s)

In light of the preceding discussion, it is evident that the talk about 'flowing' exemplifies classic emanative terminology. I would hence suggest that we take a look at what kind of picture emerges when Spinoza's position is considered from the emanativist point of view.

As I have already brought forward, Spinoza's denial of final causes has most often been considered to entail reducing all causation to efficient causation, and the way the triangle analogy has been interpreted evinces this too. For instance Carriero maintains that 'the point of the analogy is that there is no final causality in God's production of his effects: the infinite things in infinite ways that flow from the divine essence are not the consequence of some choice on God's part, directed at the good'.[27] And certainly, in Spinoza's system there is no place for a transcendent chooser-God, but this negative stand is hardly all that Spinoza wants his triangle illustration to convey. Quite the

[25] Cf.: '[E]ffect, *or* property' (defaff22exp).

[26] Note that in the *Treatise on the Emendation of the Intellect* (§ 91; C, p. 38), Spinoza talks about 'the formal character of nature [*formalitatem naturae*]' that our mind can faithfully represent; and in 5p31d he maintains that 'the Mind, insofar as it is eternal, is the adequate, *or* formal, cause of the third kind of knowledge (by 3d1)'.

[27] Carriero 1991, p. 64.

contrary: a brief overview of seventeenth-century mathematics leaves little doubt that once again, we find geometry shaping Spinoza's views.

As Paolo Mancosu explains, at the heart of the most important controversy of seventeenth-century philosophy of mathematics lay the scientific status of geometry: according to the prevailing Aristotelian conception, in order to be scientific an explanation needs to reveal the cause of the phenomenon under examination – be the relevant kind of cause formal, material, efficient, or final. Now, the question whether geometrical demonstrations were truly causal was already raised during the Renaissance period: in many cases those demonstrations did not proceed through the causes of geometrical objects and their properties. A widely discussed example of this was the way in which the sum of the internal angles of a triangle was proven to be equal to two right angles by appealing to certain auxiliary segments, which, understandably, cannot be seen as the cause of the aforementioned equality. Accordingly, this kind of demonstration did not cite causes, and so, granted the Aristotelian conception, it could not be scientific, either. Keeping geometry's paradigmatic position in mind, this was no minor issue.[28]

However, the crucial thing for us is to note what kind of causal explanations would surely accord, for the Aristotelians, with geometry's nature and could thus guarantee its scientificity:

> The scholastic tradition would have assumed this [the proof concerning a triangle's angles] to be a causal proof by maintaining *the triangle must have an essence (given by a definition) that determines, as in a formal cause, the rest of its properties*, in particular, the sum of the internal angles is equal to two right angles.[29]

In other words, a specific kind of intrinsic formal causality was seen to pertain to geometrical objects; the crux of the problem was that although the essence of a geometrical object was traditionally thought to determine the properties pertaining to the object, some important geometrical demonstrations did not refer to these essential causes. For instance, Isaac Barrow (1630–77), who delivered his lectures during the 1660s and is one of the noteworthy mathematicians Mancosu discusses, defends geometry's causality and does not hesitate to use emanative terminology; he sees mathematical propositions as flowing from the essences of things. Moreover, Barrow emphasizes formal causality as the foundation of a necessary consequence.[30]

[28] Mancosu 1996, pp. 10–15. [29] Ibid., p. 14, emphasis added.

[30] See ibid., pp. 21–2. As Mancosu (1996, pp. 17–19) points out, according to some seventeenth-century mathematicians, geometry deals not only with formal but also with material causes. To my mind, however, such passages as *Metaphysics* 1036a26–b6 (*CWA* II, p. 1636) make it rather clear that essences of mathematical objects do not include matter; see also *On Being and Essence* II (*SW*, p. 37); *CAM*

The similarity, sometimes finding its way even into the choice of words, of the foregoing geometrical discussion and Spinoza's views on causality is, I think, evident. It is simply improbable that Spinoza was not aware of the most significant mathematical debate of his time or the prevalent ways of thinking about formal causality when he decided to include the triangle analogy in 1p17s. And as the following passage quoted in the Introduction witnesses, if there is a model of causality he is led to by denying the final causes, it is much more akin to formal-essential than mechanistic causation: 'This alone, of course, would have caused the truth to be hidden from the human race to eternity, *if Mathematics, which is concerned not with ends, but only with the essences and properties of figures, had not shown men another standard of truth*' (1app; C, p. 441; G II, p. 79, emphasis added). All this, together with Spinoza's talk about formality in connection with essences and being of things, strongly suggests that Spinoza regards the mathematical standard as the correct one, because through it the true formal character of the world can be pinned down. Given Spinoza's tendency to think about *all* things through the model provided by geometrical objects,[31] it is quite understandable that his doctrine of causality has much in common with the idea of the formal cause, or 'what follows from the essence'. Consistently enough, this applies most clearly to the only substance: according to the passages concerning divine freedom, God-or-Nature has precisely as little choice over the fact that he produces everything there is in virtue of his essence alone, as fire has over its hotness, a rational creature over its ability to laughter, or a triangle over the sum of its internal angles.

To summarize: Spinoza thinks that ontology should be founded on the ideas of essence, property, and (formal-emanative) 'following', which constitute a structure most clearly seen in geometrical entities. The point is that this kind of following, in traditional parlance formal causation, has an autonomous standing and is not reducible to or to be confused with efficient causation: *it holds in geometry* even though no efficient causes (nor matter nor ends) are involved. Most importantly, for Spinoza this primary type of causality is involved in the basic structure of reality and things in it.[32] However, in the case of *real things* it is

VII.9.1468; Lear 1982, p. 169. This – if not the simple fact that Aristotle sometimes equates essence with form alone (see e.g. *Metaphysics* 1032b1–2, 1035b32; *CWA* II, pp. 1630, 1635) – probably explains why Barrow stresses the importance of the formal cause when writing about geometrical causality. As Des Chene (1996, pp. 232–5) explains, it was a debated issue in Aristotelianism whether or not matter should be included in the essence of natural things.

[31] See Ch. 1.

[32] Pasnau (2004, p. 40) writes about the formal cause in Aristotle, 'formal explanation seems to take place at a more abstract, metaphysical level. What is at issue here are not ground-level facts about why a body has this or that observable quality, but more refined questions of unity and individuation,

correct to say that *the formal character of things equals or results in efficient causality*: insofar as things are real, the essential causal architecture equals efficacy, i.e. bringing about *real* effects, states, or properties.[33] In fact, I would argue that Spinoza takes the following route to reach a conclusion rather reminiscent of the above outlined late scholastic position concerning natural things: he starts from the idea that just as properties follow from geometrical objects' essences, they follow from all essences; then, given that some of those essences belong to real existents, the properties brought about cannot but be real effects; thus, it is proper to call the essences in question efficient causes.[34] In light of this, it is understandable that once Spinoza has established that all things follow from God's nature (1p16), God of course being a real (natural) thing, the 'flowing' from his nature has the *corollary* that 'God is the efficient cause of all things' (1p16c1). In brief, Spinoza's notion of efficient causality is based on the geometry-inspired doctrine of the nature of being.

SPINOZA AND THE ESSENTIALIST MODEL OF CAUSATION

The observations made thus far suggest that Spinoza puts forward a particular essentialist model of causation in many respects similar to a traditional understanding of emanation. According to Spinoza's model, causation has fundamentally to do with the fact that as things are what they are – that is, as they have the kind of essences they do – certain properties follow or flow from those essences.[35] And since there is only one substance, God-or-Nature

requiring judgments about, for instance, a thing's modal properties.' I think the same can be said about the position of 'essential following' in Spinoza's system.

33 Michael Della Rocca (2003a, pp. 80–1) contends that 'the notion of causation somehow depends on the notion of conceivability' and asks,

> is Spinoza willing to say that any form of conceivability or conceptual connection is a kind of causality? Is there, e.g., a *causal* relation between the fact that a triangle is a right triangle and the fact that it satisfies the Pythagorean theorem? This certainly does not seem causal, but at most merely conceptual. Still, I think Spinoza would not balk at calling this relation causal.

I believe my interpretation squares quite nicely with this contention. For I argue that (efficient) causality depends on a ('conceivable') formal structure of a thing; but still a certain kind of causality may be claimed to pertain to this sort of conceivability, namely formal causality. Cf. also Matheron 1991a, p. 23; Scala 1994, p. 36; Macherey 1998, pp. 139–46.

34 So even though Spinoza's position is in harmony with the late scholastic view, as formulated in Pasnau 2004, in which the substantial form was seen as an internal efficient cause, it nevertheless should be kept in mind that for Spinoza the form of causality reigning in geometry and involving no efficient causality is the primary and autonomous one, determining the basic nature of things. Only if a thing is a real one, its intrinsic formal character can be said to be converted into 'internal efficient causation'.

35 Here I am in agreement with some noteworthy recent discussions. In his elaborate article on Spinoza's *conatus* argument, Garrett claims Spinoza to have endorsed a view according to which all finite things inhere in God, and since any y that inheres in x is both conceived through and caused by the essence of x (Garrett 2002, pp. 136–42, 144–5), this represents one essentialist way of understanding Spinoza's

that is also a real thing, indeed *ens realissimum*, it is understandable that everything turns out to be what it is and the way it is because God's essential causal activity results in real effects, or 'infinitely many things in infinite modes'.

At least the following points can be said in support of the view that the best way to make sense of Spinoza's conception of causation is by reading certain key passages against the background in which the idea of formal-emanative causation holds a prominent place.

First and foremost, 1p16 seems to be talking about the sort of production of necessary properties characteristic of the formal cause. Aquinas is following the tradition when he states in his *On Being and Essence*, 'a thing is intelligible only through its definition and essence' (*SW*, p. 35), and as we have seen, Spinoza obviously adopts this contention without hesitation: each thing has its definable essence. The triangle analogy of 1p17s is there to drive home the idea that certain properties belonging to any genuine thing follow from its definable essence with the precisely same kind of necessity as in geometry. And on these grounds we are in the position to know that if such an entity is real, it has efficacy, in virtue of its essence, to produce those properties. This observation reveals that all things, be they mere beings of reason or real things, share the same formal architecture of 'following', which explains, at least in part, Spinoza's alleged disregard of the distinction between logico-geometrical following and efficient causation: at least in God's case, the latter occurs as determined by the former, thereby necessarily realizing everything that follows from God's essence. This bears, I think, a striking resemblance to the emanativist way of speaking about both geometrical following and the immanent causal activity of physical and mental things in terms of emanative production of properties. Moreover,

notion of causation. In consonance with this, Garrett emphasizes the causal efficacy of essences: 'In Spinoza's view, something is an individual thing only to the extent that it has some nature or essence through whose genuine activity effects can be understood to follow' (Garrett 1999, p. 330). 'For Spinoza, an individual or singular thing exists to the extent that there is instantiated a definite essence or nature that can serve as a locus of causal activity. Where there is such an essence, properties follow (both causally and logically) from that essence' (Garrett 2002, p. 150; see also 1991, pp. 194, 201). In his article defending a teleological reading of Spinoza, Martin Lin (2006, p. 343) claims similarly that, for Spinoza, 'things are causally efficacious only in virtue of their essences'; but Lin links this kind of 'causation through essence' with a Neoplatonic view of efficient causation according to which 'efficient causation is a kind of giving' in which the cause resembles the effect (see also Lin 2004, pp. 29–33). Although Garrett and Lin are right in emphasizing the importance of essences in Spinoza's theory of causation, I think that the formal-emanative framework allows us an even better grasp of the import and nature of Spinoza's essentialist commitments; for instance, Spinoza is pushed towards his essentialist view on causation not only, as Lin (2006, p. 345) claims, by his 'rationalistic cast of mind' that shuns brute contingencies, but by his *geometrical* brand of rationalism and the model of causation it carries within.

the fact that geometrical objects were traditionally also regarded as causal entities helps to understand why Spinoza finds it so unproblematic to make the transition from the basic ontologico-conceptual considerations (i.e. definable essences and the being determined by them, as in 1p16 and 1p16d) to the causal realm (things as causes in virtue of their essences, 1p16c1).

Second, although the term 'emanation' itself is not to be found in the *Ethics*, Spinoza uses it in his correspondence, for instance in the following fashion:

> [W]hether the good that follows from virtue and the divine love is bestowed on us by God as judge, or whether it emanates from the necessity of the divine nature [*ex necessitate Divinae naturae emanet*], it will not on that account be more or less desirable. (*Ep*75; S, p. 945; G IV, p. 312)[36]

Of course, Spinoza himself endorses the latter of these two views.

Third, as Gueroult explains, the passages of the *Ethics* that deal with causation (1p16–p18) place into the Spinozistic framework a considerable number of the different ways in which Franco Burgersdijk and Adrian Heereboord (two seventeenth-century philosophers from Leiden with whose work Spinoza was acquainted) classify causes; and in this context 1p16 refers precisely to the emanative (vs. active) cause.[37]

Now, if Spinoza's conception of causality really is as close to the formal-emanative one as I claim it is, we should be as clear as possible about its relation to other types of causes. As I mentioned earlier, the final and the material cause pose no problem here: at least with regard to God, Spinoza denies that there is any kind of teleology, and the material cause seems to be altogether expunged from his thought. This leaves us with the efficient cause; Spinoza undoubtedly wants to include it in his system, and now the question is, how is efficient causation accounted for within the essential model? This is a thorny question to say the least, but I believe we have already gathered enough material to answer it. Textual evidence both for the 'emanative-essentialist' and the 'efficient-mechanist' readings of Spinoza's thought can be found; obviously, he regarded the two aspects as quite compatible. Nowhere, to my mind at least, is this witnessed more clearly than in the following already cited passage: '[N]othing belongs to the nature of anything except what follows from the necessity of the nature of the efficient cause' (4pr; C, p. 545; G II, p. 208). This can be taken as a piece of

36 See also *Ep*43.

37 Gueroult 1968, pp. 246–8, 251–6. As Gueroult (p. 297) notes, according to Burgersdijk form is the emanative cause of properties.

evidence for the efficient reading;[38] but it can be read the other way round, too, as saying that real things act as efficient causes, and do so because effects necessarily follow from their natures or essences. Actually, this way of reading it concords much better with Spinoza's order of exposition in 1p16 and its corollary, where efficient causality is a consequence of essential causation, and 3p7d, where he states, '[f]rom the given essence of each thing some things necessarily follow (by 1p36), and things are able [to produce] nothing but what follows necessarily from their determinate nature (by 1p29)'. Real things' essence-originating causing can be called efficient because it results in real changes in the agent and other things – indeed, by what other term could this aspect of causality be characterized? However, the key idea is that without the essence-originating, formally structured causal thrust there would be no efficacy in the first place.[39]

The flowing or following from God's nature results in real effects which realize a specific ontological structure; this is something that cannot be affected or intervened in for the simple reason that, in the Spinozistic framework, there is nothing besides God. Because finite things are among the real effects of God's productive activity, God can be called their efficient cause. However, the causality of finite things, or of modifications of God, is another issue, for, unlike God, they are not causally isolated, exclusively self-determining agents. As a consequence, only God's causality amounts to 'full-blown' emanation, which must be taken into account when the essentialist model is used to explicate the causation taking place between finite things.[40]

[38] Carriero (2005, p. 130) writes: 'By "the necessity of the nature of the efficient cause" I take Spinoza to mean what I have referred to as a blind efficient cause.'

[39] Also Lin's (2006) position is, I believe, close to the one presented here. Moreover, it should be noted that Spinoza's views have here, too, some noteworthy affinities with the late scholastic ones: as Hattab observes, the late scholastic camps debating the nature of efficient causation taking place between substances agreed that the efficient cause–effect relation is grounded in some feature (intrinsic mode) either of the agent or the patient (Hattab 2003, pp. 7–8). Now given that in this tradition it is the essence that determines what the substance is and what kind of properties, modes, or features it necessarily has, I think it can be said that efficient causation has its basis in formal (or emanative) causation. Spinoza, in turn, holds there to be things with essences, from which certain things follow and based on which things determine each others' actions, this determining equalling transeunt efficient causation (more on this below).

[40] However, it should be kept firmly in mind that since there is, according to Spinoza, only one substance, considering the essentialist model from the viewpoint of finite things should not be done as if finite things were separate substances (this is what happens quite quickly due to our tendency to endorse a pluralistic ontology). After all, Spinoza thought the only metaphysically appropriate point of view to be the one in which all causation is immanent to the monistic substance (see 1p18); obviously, all talk about causal relations taking place between finite things can and should be translated, as it were, into more adequate talk about God's internal causation (indeed, this is why

Because finite things enter into relationships of interdetermination with each other,[41] they all are under the influence of external causes – whose fundamental manner of operation is respectively determined by their individual essences. These interactions always determine, at least partly, the way a finite thing behaves. In other words, two cases must be distinguished: that in which a thing causes something in virtue of its essence alone, and that in which the thing's essence is only a partial cause of the resulting effect (which means that something happens that is a joint product of two or more disparate things' essences).[42] In this respect the difference between God and his modifications lies in the fact that God's causal activity is determined by his essence alone, whereas finite things' causal activity is determined by other finite things, i.e. by external causes, too.

I think Spinoza attempts to give an uncomplicated formulation to issues pertaining to the necessity of causation by saying that '[f]rom a given *determinate* cause the effect follows necessarily' (1a3, emphasis added): regardless whether an agent is determined to causing effects internally (i.e. actively) by its own nature alone, or externally (i.e. passively) so that the resulting effects also depend on the nature of external causes, once the cause is determined, the effect results with the necessity pertaining to the emanative cause. Moreover, I would like to emphasize that in the case of passivity, too, we are still dealing with what follows from the essences of things (the agent and the patient), or what is the same, joint causing of effects is also, for Spinoza, essence-based causing.[43] So although Spinoza uses the famous example of a stone that 'receives from the impulsion of an external cause a fixed quantity of motion whereby it will necessarily continue to move when the impulsion of the external cause has ceased' to illustrate that 'every single thing is necessarily determined by an external cause to exist and to act in a fixed and determinate way' (*Ep*58; S, p. 909; G IV, p. 266), this should not be taken to mean that the resulting motion would be what it is were the essence of the stone different from what it is; the same impulsion would not, of course, effect the same kind of motion if directed to, say, a feather. This idea is clearly stated by the following axiom:[44]

the Suárezian account of emanation, explicating the way in which a substance obtains its properties, sits so well with Spinoza's thought). As Koistinen (2002, p. 60) points out, Spinoza thought that his monistic doctrine is capable of unlocking the severe problems plaguing both Descartes's interactionism and Malebranche's occasionalism. On immanent causation in Spinoza's system, see especially Nadler 2008; but also Lennon 2005, p. 28; Sévérac 2005, p. 56; Lærke 2009, pp. 183–7.

[41] See the previous chapter, especially the discussion on formal being.

[42] According to 3d2, the first case equals activity, the second passivity; see Ch. 3 below.

[43] For a more detailed explication of this, see Ch. 6 below.

[44] For another illustration of this idea, see *TP* IV.4; S, p. 697; G III, p. 293.

> All modes by which a body is affected by another body follow both from the nature of the body affected and at the same time from the nature of the affecting body, so that one and the same body may be moved differently according to differences in the nature of the bodies moving it. And conversely, different bodies may be moved differently by one and the same body. (2le3a1)

It should be noted, however, that although any finite thing is capable of operating in virtue of its essence, it can itself neither determine the kind of essence it has nor come into existence because of that essence, for all finite things come to exist from external causes (1p11s).[45] This means that from the Spinozistically adequate monistic standpoint finite things are God's states which God produces via his other states,[46] so although finite things can be said to be essentialist causers, they are brought about by other finite things, which, operating according to their essences, act as efficient causes on other finite things; and as 1p28 states, this chain of finite causes continues *ad infinitum*. But we should not take Spinoza's idea to be that all finite things are always *completely* externally determined; as the axiom above and Spinoza's view of passivity imply, and as Spinoza elsewhere makes clear,[47] things are also active and self-determined in different degrees.[48]

At this point I would like to address a possible objection to my interpretation. It has been claimed,[49] largely on the basis of the influential reading Curley presents in his *Spinoza's Metaphysics* of 1969, that the general features of the universe that are explicable by laws of nature can be said to emanate from God, but not singular things each one of them a part of, and determined by, an infinite causal chain that consists of finite things. In a way similar to the covering law model of Carl Hempel and Paul Oppenheim,[50] Curley contends that neither infinite modes (the general nomological facts) nor finite modes (the singular facts)

> are by themselves adequate causes of finite modes. Taken separately, they are only partial causes; the existence and actions of a particular finite mode cannot be understood either by reference to other finite modes alone or by reference to infinite modes alone, but only by reference to both infinite and finite modes.[51]

So, 'the infinite and finite modes are separately necessary and only jointly sufficient conditions of finite modes'.[52] According to a line of thought

[45] For more on this, see the next chapter.
[46] As Koistinen (2002, p. 68) explains, following Roderick Chisholm, 'individuals enter into causal relations *via* their states'.
[47] See e.g. 2p29, defaff1; *TTP* III (S, p. 417; G III, p. 46).
[48] For more on this, see Ch. 5 below.
[49] See Rice 1992; for relevant discussion, see also e.g. Watt 1972; Lennox 1976; Gilead 1990.
[50] See Hempel and Oppenheim 1948.
[51] Curley 1969, p. 66.
[52] Ibid., p. 70. See also Curley 1988, pp. 47–50.

inspired by this, a distinction can be made between two kinds of causality in Spinoza's system: the *emanative* one pertaining to (immediate and mediate) infinite modes, and the *sequential* one pertaining to finite modes.[53]

However, despite its merits,[54] this line of interpretation is not without its problems. The weightiest one is that if only general facts are claimed to follow from God's nature, is seems that there is and cannot be any explanation for why precisely the actual sequence of finite modes exists – or for why there are any finite things in the first place.[55] When confronted with this dilemma,[56] Curley argues (with Gregory Walski) that 'the existence of the totality of finite things and [. . .] the existence of the most general laws governing finite things' simply are by their very nature such phenomena that there can never be a proper (scientific) explanation for them; and an explanation should not be sought when it is even in principle impossible to find one.[57] However, as we saw in the previous chapter, Spinoza obviously holds that there *is* an ultimate explanation for everything, namely God's nature, from which all things follow so that a specific immutable and eternal ontological structure is realized.[58]

[53] Rice 1992, pp. 46, 48–50.

[54] As Diane Steinberg (2000, p. 25) comments, Curley's interpretation 'gives us a relatively concrete understanding of the mysterious infinite modes and explains how they are caused by the absolute nature of an attribute'.

[55] Cf. Huenemann 1999, p. 225.

[56] The objection is by Huenemann and presented in Curley and Walski 1999, p. 258 as follows: '"[Y]our [Curley and Walski's] Spinoza cannot explain why one possible universe is actual and another is not. That this universe, and no other is actual is, I take it, a brute fact, true independently of any fact about God."' Huenemann articulates well what seems to be the greatest problem in Curley's interpretation:

> Monism has been understood generally as the view that somehow all things [. . .] owe their existence and essence ultimately and exclusively to a single thing, and Spinoza has been understood generally as such a monist. But according to Curley, Spinoza does not claim that finite modes owe their existence *exclusively* to God. God's nature plays a part in their existence, to be sure: it grounds the laws of nature, which are necessary for the generation of any finite mode. But the laws alone are not sufficient for a finite mode's generation. There must also be an independent causal chain of finite modes, a chain whose links are again governed by the laws of nature, but which (again) do not owe their existence exclusively to God. [. . .] [T]he totality of finite modes cannot be understood as following from God's nature or God's attributes or God's infinite modes, according to Curley, since this would mean that each finite mode does so, and E IPP21–23, 28 rule this out. Thus there is no final explanation, in Curley's account, for the existence of the totality of finite modes other than the individual explanations for the existences of particular members of that totality. (Huenemann 1999, p. 227)

However, discussing Huenemann's (1999, pp. 235–8) own solution to these problems would take us too far afield.

[57] Curley and Walski 1999, p. 259.

[58] Huenemann (1999, pp. 227–8) mentions, justifiably I think, also *KV* 1.4 (C, pp. 83–4; G 1, p. 39), *CM* 1.3 (C, p. 309; G 1, p. 243), and 1p29 as additional reasons to see Spinoza as believing, as Huenemann puts it, 'that some ultimate explanation [. . .] must be given for the existence and determinate nature of the totality of finite modes' (p. 228).

As we will see later on,[59] Spinoza's rather strong brand of geometrical essentialism lurks behind his general ethical project, moral psychology, and political philosophy, too, and has direct implications for how human affects are conceived. Indeed, before embarking upon constructing his theory of the affects, Spinoza proclaims to 'consider human actions and appetites just as if it were a Question of lines, planes, and bodies' (3pr; C, p. 492; G II, p. 138), thus leaving little doubt that he understands his approach to be geometrical, not mechanistic in character. And such propositions as '[e]ach affect of each individual differs from the affect of another as much as the essence of the one from the essence of the other' (3p57) exemplify his essentialist psychology.[60] Also, the doctrine of our actual essence as *conatus*, central for Spinoza's psychology, witnesses the fact that proper grasp of his essentialism is necessary for the correct understanding of his naturalistic ethics. In brief, appreciating the geometry-inspired model of causality is an indispensable step towards understanding Spinoza's geometrical dynamics of human existence.

The longstanding dominance of the Humean conception of causality may make Spinoza's view appear, at first sight at least, strange. However, the recent developments in contemporary metaphysics and philosophy of science, sharing a host of the essentialist intuitions behind Spinoza's thought, may make him easier to approach. To take an important present-day representative of this line of thought, Brian Ellis holds that what things 'do or could do is of their essence'.[61] Contentions such as this are part of an on-going discussion the direction of which is in the making, but these developments at least suggest that Don Garrett's quite positive assessment of Spinoza's fortunes is well-founded: for Garrett's Spinoza, 'thinghood is a function of manifesting a nature suitable for playing a substantive role in explanations – that is, having a nature from which things follow causally and through which they can be understood' – and this is a position which Garrett estimates is 'attractive and deserving of serious consideration'.[62]

To sum up, I think it can be said that Spinoza discards the Aristotelian doctrines of teleology and of activity as actualization of potentiality, but not the geometry-inspired essentialism stemming from that same source: from

[59] See esp. Ch. 6.

[60] For the impact of essentialism on Spinoza's political theory, see *TP* IV.4 (S, p. 697; G III, p. 293); *TTP* XVII (S, p. 536; G III, p. 201).

[61] Ellis 2002, p. 1. Already Rom Harré (1970, p. 88) claims that we may see the behaviour of things 'as flowing from their natures or constitutions as consequences of what they *are*. So they must behave in the specified way, or not be the things that they are.' This essentialist line of thinking is developed more extensively in Harré and Madden 1975.

[62] Garrett 2002, p. 151.

this viewpoint, causality is not about regular succession of event types but about finite things with essences in virtue of which they produce effects and determine each others' manner of acting. Once a thing is determined to act in a certain way, be that determination brought about inwardly or externally, the effect necessarily results. However, when dealing with finite things' causation, we should not lose sight of Spinoza's original monistic vision – as we have seen, from the philosophically adequate eternal standpoint there is, in the end, only one substantial causal agent, God-or-Nature, who, when producing everything there is, determines the nature of reality with unyielding rigour: '[A]ll things have been determined from the necessity of the divine nature, not only to exist, but to exist in a certain way, and to produce effects in a certain way' (1p29d).

CHAPTER 3

Power, existence, activity

POWER IN LATE SCHOLASTIC AND CARTESIAN METAPHYSICS

We are finally in a position to move into discussing the focal notion of this study, that of power. There is a story to be told about the concept's fate in late medieval and early modern thought, a story which helps us to understand what is at stake when Spinoza introduces the concept into his system. Of course, scholastic discussions on power and causality are extremely complex, but I believe the following general observations to be accurate.

In medieval Aristotelian metaphysics, it went unquestioned that natural things are causally efficacious.[1] In fact, this was commonly seen as something experience straightforwardly proves, '[f]or what is better known to the senses than that the sun gives light, fire produces heat, water cools?' (*MD* 18.1.6). Indeed, as Des Chene comments, truths like the ones Suárez mentions here are, for the Aristotelian, 'so obvious that merely to point out that a doctrine would have the effect of denying them was sufficient to prove its absurdity'.[2]

Causal efficacy was, in turn, explicated in terms of intrinsic powers of things. The basis for this can be found in Aristotle's *Metaphysics*, where the first discussed way to use the notion of power (*dunamis*) is to mean by it 'the source, in general, of change or movement in another thing or in the same thing *qua* other, and also the source of a thing's being moved by another thing or by itself *qua* other' (1019a18–20; *CWA* II, p. 1609).[3] Now, Aristotelian natural things are hylomorphic entities consisting of matter and form, and since the basis of powers lies in the latter, I will begin by taking a look at the relationship between (substantial) form and power.

[1] In *MD* 18.1.5, Suárez offers a particularly revealing account of this. As Alfred Freddoso (1988, p. 99) puts it, '[t]he medieval Aristotelians exhibit little patience with those who espouse occasionalism or theories closely resembling it'.

[2] Des Chene 1996, p. 21.

[3] Translation by W. D. Ross.

As we have seen, there was some disagreement among the scholastics over the nature of emanative causation (whether or not it is a real action), but here it is crucial to appreciate what was universally agreed on: active powers count among the proper accidents that somehow emanate from or pertain to substantial form. Moreover, forms act or cause effects on other things *only* through their powers.[4] As Alfred Freddoso explains, from the natures of things

> flow the intrinsic causal powers, tendencies, and dispositions that are definitive of the various natural kinds and that are thus possessed by the individual instances of those natural kinds. So it is metaphysically necessary that, say, tomato plants tend to produce tomatoes and that fire tends to incinerate human flesh brought into close proximity to it.[5]

Although dealing with the mind and its operations, this line of reasoning seems to be at work in an oft-cited passage of the *Summa Theologica* where Aquinas delineates the Aristotelian position as follows:

> The powers of the soul are its natural properties. But the subject is the cause of its proper accidents; whence also it is included in the definition of accident [. . .] Therefore the powers of the soul proceed from its essence as their cause. (*ST* I, Q. 77, A.6)

In his explanation of the issue Aquinas notes, 'all the powers of the soul [. . .] flow from the essence of the soul, as from their principle' (*ST* I, Q. 77, A.6). And as we have seen, for Suárez there is a strong (and causal) connection between the substantial form and certain ways of behaving characteristic of natural agents.[6] So it seems that generally for the scholastics, the substantial form is not only a crucial ingredient in individuating things but also something from which certain powers 'flow', and through these powers effects are caused. Moreover, to each thing various powers belong,[7] and the substantial form, distinct from these powers, functions as their unifying ground.[8] We should note the empiricism involved in this view: that there is such an entity as the substantial form, with the efficacy to bring about accidents some of which are powerful, can be known through our senses; we have a sensory idea of the effects the best explanation of which is the substantial form.[9]

[4] Des Chene 1996, pp. 157–61. [5] Freddoso 1988, p. 109. See also Des Chene 1996, pp. 216–17.

[6] Cf. Des Chene 1996, pp. 71–5, 157–61; Hattab 2009, pp. 45–7.

[7] Max Jammer (1999 [1957], p. 35) lists heat, cold, chemical activity, botanical functions, hardness, and light as things that the Greeks spoke of as powers.

[8] This is so especially for Suárez; see Des Chene 1996, pp. 73, 158, 161; Hattab 2009, p. 49.

[9] See esp. Hattab 2009, pp. 45–6, 148; but also Secada 2000, p. 11.

In the Peripatetic framework natural changes are, of course, understood as end-governed actualizations of potentialities, and this has an effect on how powers are conceived: they are *potential capacities for change* that, once exercised, end in something actual.[10] To take a classic example, a seed has the power or potential to become a tree; if placed in suitable circumstances, the exercise of that power begins and becomes manifested as an actual tree. The following passage illustrates the Thomist line of thought in these matters:

> Again, it is absurd to say that a body is not active because accidents do not pass from one subject to another. For when we say that a hot body gives heat, we do not mean that the identical heat which is in the heater passes into the heated body: but that by virtue of the heat in the heater, another heat, individually distinct, becomes actual in the heated body, having been potentially therein before. Because the natural agent does not transmit its own form into another subject, but reduces the passive subject from potentiality to act. (*SCG* III.69)

As Eileen O'Neill holds, this represents a particular view of causation as eduction, according to which natural change is 'a process of act coming to be out of potency' which means that 'nothing passes over to the recipient' but that '[a] body brings about a change in another body, or in a substantially united composite of body and soul, by educing it out of the patient's own potentiality'.[11] In this kind of framework, things cannot begin to exercise their powers all by themselves: another power as an efficient cause is needed for this to happen.

I believe that these observations suffice for our purposes and later will help us in interpreting Spinoza's thoughts on power. At this stage I want to draw attention to two things. First, the linkage between nature, causality, and power can be considered a truly time-honoured one; that Spinoza rejects certain key elements of the just-sketched view should not keep us from appreciating the importance this constellation of concepts has for him. Second, medieval Aristotelians can justifiably be said to have held that any true natural thing must pass the Eleatic Stranger's reality test, or as Freddoso succinctly puts it, '[t]o Aquinas, Molina, Suárez, or any other robust Aristotelian, denying active causal power to an entity amounts to nothing less than denying that entity the status of being a substance'.[12]

In the reshaped early modern intellectual landscape, the concept of power was conceived in a considerably different way. As so often, Descartes emerges here as Spinoza's most important seventeenth-century

[10] See Des Chene 1996, pp. 23, 29–30. [11] O'Neill 1993, p. 38; see also pp. 39–40.
[12] Freddoso 1988, p. 109.

predecessor.[13] Greatly influenced by the new physics as he was, Descartes uses not only *potentia* (or *puissance*) but also *vis* (*force*), a notion familiar from the new physics,[14] without making explicit how he understands the relationship between the two notions. The Cartesian God's power will be discussed below; but here I would like to underscore the way in which Descartes's mechanics-inspired view on material things' power differs from the Aristotelian one: in the *Principles of Philosophy* (henceforth *PP*), he considers any body to have a force (*vis*) understood as a tendency to persist in the prevailing state (*PP* 2.43; CSM 1, pp. 243–4), and this force can be measured in terms of speed and size.[15] Hence, Descartes shows no need to refer to anything even remotely resembling substantial forms or actualization of potentialities.

Descartes's views on power have stirred much debate ever since their inception and continue to do so even today. The reason for this is a tension located right at the heart of his philosophical project. Now, Descartes famously writes:

> [E]ach substance has one principal property which constitutes its nature and essence, and to which all its other properties are referred. Thus extension in length, breadth and depth constitutes the nature of corporeal substance [. . .] Everything else which can be attributed to body presupposes extension, and is merely a mode of an extended thing [. . .] For example, shape is unintelligible except in an extended thing; and motion is unintelligible except as motion in an extended space. (*PP* 1.53; CSM 1, pp. 210–11)

So extension is the essence of corporeal entities, and other bodily properties are modifications of extension. But this does not prevent Descartes from appealing to forces and causal powers, which, in turn, generates what may be called the Cartesian problem of power.[16] I think the following recent statement captures the crux of the issue:

> Since matter can have no qualities other than the attributes shared by all substances (such as existence and duration) and the modes of extended substance (such as size, shape and position), attributing causal powers to nature explicitly contradicts Descartes's notion of matter as passive extended substance.[17]

[13] For Hobbes's views on endeavour and power, see Ch. 4.

[14] For Galileo and force, see Jammer 1999 (1957), pp. 94–103.

[15] See *PP* 2.36, 2.43 (CSM 1, pp. 240, 243–4); see also Garber 1994, pp. 44–5; Des Chene 1996, p. 340.

[16] Cf.: 'But force, although it has no obvious place among the modes of extension, cannot, it would seem, be dismissed entirely in the way that he [Descartes] dismisses weight or occult qualities. Call this the *problem of force*' (Des Chene 1996, p. 313).

[17] Hattab 2007, p. 49.

So the dilemma is, how can any kind of power, force, or activity be assigned to natural agents, if their essence is mere (inert) extension?

It is uncontroversial that when facing this problem, Descartes's main strategy is to attempt to ground the powers of material things in the active power of God. The famous claim is that 'God is the primary cause of motion' who 'imparted various motions to the parts of matter when he first created them, and he now preserves all this matter in the same way, and by the same process by which he originally created it' (*PP* 2.36; CSM 1, p. 240). But instead of solving the problem this, of course, only pushes it a step further, for it is far from clear how the relation between God's activity and the power of bodies should be conceived. Moreover, as we will see in more detail when mapping the background of Spinoza's *conatus* doctrine, there is also another, less conspicuous strategy Descartes devises in order to alleviate the problem of power: as we saw, he identifies power of material things with the *tendency* to continue *ad infinitum* in the present motion, or more generally in the prevailing state, unless interfered with by external causes; and Descartes seems to hope that this kind of tendency-like power would offer him what his physics requires while being ontologically harmless enough to be reconciled with the thesis that extension constitutes the essence of bodies. But as I will argue later on, even *if* it turned out that this line of thought could in some sense solve the problem of power, Descartes would not be out of trouble, for arguing in this way leads to a new problem, namely that of resistance.

I do not wish to take a definite stand on how the problem of power and its outgrowth, the problem of the relation between God's and bodies' power, should, in the end, be viewed.[18] I would only like to point out that here Descartes appears to want his cake and eat it too: his aspiration to expunge everything 'occult' from his philosophy leads him to define the nature of matter as extension; but this, in turn, makes a notion he wants to have in his system, that of power, potentially unavailable for him, and its

[18] The issue is difficult and discussions complex; but according to Gueroult (1980, esp. pp. 196–8), Alan Gabbey (1980, pp. 234–5), Tad Schmaltz (2008, pp. 116–21), and Paul Hoffman (2009, esp. pp. 127–32), Cartesian bodies do have power of their own in some genuine sense, whereas Jammer (1999 [1957], pp. 103–5), Gary Hatfield (1979), Daniel Garber (1992a, pp. 293–305, 1992b, pp. 320–2, 2001, chs. 8–10), and Des Chene (1996, pp. 277–8, 312–44) defend contrary views. Della Rocca's (1999, pp. 58–60) position seems to occupy a middle ground: bodies purely geometrical by their nature have force in the sense that objects stand in a certain relation to God. And it surely looks as if, for Descartes, there is force or power in bodies only insofar as they are considered in relation to God. Thus, 'the concurrentist reading' – put forward by such scholars as Andrew Pessin (2003) and Hattab (2007) – has clear strengths (see also Pietarinen 2009a, p. 156), and Descartes does say that God 'concurs' (in Latin, *concurrere*, literally 'to run together') with the causality of finite things. However, I am unsure of the extent to which this solves the problems involved.

meaning rather vague, which is testified by the occasionalistic developments and the contesting contemporary interpretations of his thought. It suffices for my purposes to locate what seems to be the evident source of the irresoluteness plaguing Descartes's position: he severs the pivotal connection between essence and power.[19] But when power is regarded as something things have by their nature, these problems do not arise, or at least there seem to be very promising solutions readily at hand. If things are seen, in essence, as causally powerful, speaking of power is wholly appropriate, and there are good chances of finding a proper grounding for the idea of resistance to opposition.

With this context and Spinoza's essentialist model of causation clarified, we are in a position to outline the meaning of Spinoza's notion of power and to discover his reasons for including it in his system.

POWER REHABILITATED BY SPINOZA

The opening part of the *Ethics*, where Spinoza lays the foundations of his philosophy, is titled 'On God' (*De Deo*). The reason for this is, of course, that the only substance is there identified with God. To call substance, i.e. the whole of nature, God may be regarded as superfluous, or worse still, blasphemous, but I agree with those who regard Spinoza as taking seriously the idea that Nature is God. In Bennett's words:

> Spinoza's main reason for liking 'God' as a name for the entire natural world is that the world comes closer than anything else to fitting the traditional Judaeo-Christian account of God. If God is to be infinite, eternal, not acted on by anything else, the ultimate source of the explanation of everything, and not susceptible to criticism by any valid standard, then God must be Nature as a whole; or so Spinoza thinks.[20]

Spinoza's substance is, for him, the only thing that meets the criteria of divinity. As is well known, Spinoza's philosophical theology differs radically from traditional ones; by giving his own account of such divine attributes as infinity, simplicity, eternity, activity, and so on he may be said to attempt to reveal the philosophically adequate meaning of traditional theological notions. In addition, the concept this study focuses on also figures here; one of Spinoza's purposes is to explicate what another traditional notion, that of *omnipotence*, really means.

[19] Cf. Des Chene 1996, pp. 323–4. [20] Bennett 1984, p. 33.

Halfway through the opening part of the *Ethics*, Spinoza claims to have shown that there is only one substance, namely God (1p14), and that this substance exists necessarily (1p11). Moreover, as we have seen, everything derivable from God's definition is also brought about by his essence (1p16). Now, supposing that all this holds, it may and should be asked, what kind of being is this entity, the only substance, that has been proven to be the cause not only of itself but of everything existent? When Spinoza further describes the essence of the being that is the cause of itself and of all things, we find him opting for the concept of power. The crucial characterization is located in 1p34: 'God's power [*potentia*] is his essence itself.' This proposition is the cornerstone of Spinoza's philosophy of power, making its presence felt especially through the so-called *conatus* proposition (3p6), so a firm grasp of its meaning is required.

The identity stated in 1p34 is not unprecedented in Spinoza's writings: it can be found already in *Descartes' 'Principles of Philosophy'* where Spinoza notes, after using the two terms interchangeably, that 'the power by which the substance preserves itself is nothing but its essence, and differs from it only in name' (*PPC* 1p7s; C, p. 250; G I, p. 163). In the *Ethics*, the unargued-for anticipation of 1p34 in 1p17s, 'God's supreme power, *or* infinite nature', suggests that he sees this identification as quite natural.[21] Now, we have already seen that the concept of power, much used in scholastic philosophy, was a problematic one for an early modern thinker such as Descartes, impressed as he was by the advances of the mechanical sciences. According to the mechanistic view, scientific explanations should be made in terms of the motion of particles of matter that form relatively stable structures,[22] and offering empty *virtus dormitiva* type of pseudo-explanations should be avoided altogether.[23] But before accusing Spinoza of being guilty of obscurantism in the sense of presenting an occult quality of power he should rather be trying to eliminate, we need to look at the reasons he gives for 1p34, stated in that proposition's demonstration:

[21] In many places, Spinoza indicates that the identity of power and essence is shown by 3p7, '[t]he striving by which each thing strives to persevere in its being is nothing but the actual essence of the thing'; see 4p33d, 4p53d; *Ep*64. But 3p7 has (via 3p6d) 1p34 in its ancestry, so 1p34 is obviously the fundamental proposition. We may yet note that, as Wolfson (1961 [1934] I, pp. 400–5, 421) explains, during the medieval period God's causality was characterized by such attributes as intellect, will, and power. But – and this reflects the importance the concept has for Spinoza – in the *Ethics* only power is identified with essence, whereas intellect and will are merely modes of substance. *Metaphysical Thoughts* (II.8; C, p. 330) still equates intellect, will, and power with each other.

[22] Nadler 1998, p. 520.

[23] As Curley (1988, p. 115) explains, in this kind of explanation '[t]he cause is identified only in terms of the kind of effect it has'; this is what happens when, for instance, opium is said to put people to sleep because of its dormitive virtue, which, in turn, is simply the power to put people to sleep.

For from the necessity alone of God's essence it follows that God is the cause of himself (by p11) and (by p16 and p16c) of all things. Therefore, God's power, by which he and all things are and act, is his essence itself, q.e.d. (1p34d)

So, to simplify matters slightly, the main line of thought in 1p34 is that since God's essence is the cause of everything, Spinoza claims it to be power.[24] Now, what is the rationale behind this? I would suggest that the identification of causally efficacious essences and power has to do with the fact that from 1p16 onwards Spinoza discusses the nature of causality, and he clearly finds 'power' to be a suitable term for characterizing it;[25] talk about causality can be translated, as it were, into talk about power. This is by no means an eccentric linkage: even today 'power' is commonly taken to refer to causal capacities,[26] and as we have seen, this has been so at least since the heyday of scholasticism.

All this is still relatively rudimentary; to acquire a better grasp of what is involved here, we should give an account of the way in which power figures in the above presented geometrized essentialism. If power is identified with the essence of God, could this not be understood as an attempt to reduce the notion of power to those of essence and causation?[27] Now, although the phrasing of 1p34 might tempt one to reply affirmatively, I believe this temptation should be resisted. True, it is only after the concept of power resurfaces in the absolutely focal 3p6 that dynamic concepts start to carry considerable metaphysical weight; but we should not overlook the way in which Spinoza has already put the concept to use in the second alternative proof of God's existence (1p11d3), or the way he invokes it only two propositions after 1p34 when he argues that all essences, including those of finite things, are causally efficacious. So although much of what is said in *De Deo* is and arguably can be stated without reference to power, in 1p34 Spinoza seems not to be trying to dispose of the concept as much as he is attempting to restore its traditional connection to the notion of nature by redefining it in a clear, non-occult manner. This, in turn, is something that placing the notion in the geometry-inspired metaphysics is supposed to

[24] For the close connection between power and causality, see also *TTP* I, III (S, pp. 403, 417; G III, pp. 28, 46). Cf. also Lin 2004, p. 38.

[25] That Spinoza considers the connection between power and causality self-evident is already expressed in the *Metaphysical Thoughts*: he claims thought to be 'a power of doing each one, of affirming and denying' and by this power 'of course, nothing else can be understood than a cause sufficient for each one' (*CM* II.12; C, p. 346; G I, p. 280).

[26] As Bennett (1984, p. 74) puts it, power is a 'paradigmatically causal notion'. This is reflected by the contemporary contention that '[t]he general notion of power involves the capacity to produce or prevent change' (Green 1998, p. 610).

[27] For a reading that understands 1p34 along such reductionist lines, see Gueroult 1968, pp. 376–82.

enable him to do. So what, then, does Spinoza mean by the power of God? And what is the reasoning behind introducing the concept in the opening part of the *Ethics*?

Given that by 'power' Spinoza refers to the intrinsic causal activity of things, and that things are what they are and the way they are because everything there is follows, with geometrical necessity, from God's essence – whereby is constituted an intelligible, strictly determined, and eternal world – the following picture emerges. Each thing has its definition, just as geometrical objects do, from which certain properties can be inferred; in God's case these properties equal everything possible. But unlike geometrical objects that are mere beings of reason (*entia rationis*), God is a *real* thing, indeed the most real thing there is (*ens realissimum*)[28] – showing that the absolutely infinite being (1d6) necessarily really exists is, of course, the point of the proofs Spinoza puts forward in 1p11. There is nothing outside God (1p15), and because no thing can exist without its necessary properties, we arrive at 1p16 which asserts that God necessarily brings about all the properties inferable from his definition (which, in fact, means that God realizes all genuine possibilities). Now, precisely at this point – and *no earlier* – the notion of power steps in: *the realization of this necessary system of entities requires power.*[29] In other words, because God is a real being endowed with causal power, God himself and all of his properties (i.e. modifications) come to be realized.[30] About geometrical objects we can say that certain properties are derivable from their definitions and follow from their essences (which function as kinds of formal causes), but as they are not real entities, they do not have power to realize their properties as real effects. Thus, *real being requires power* even at the most fundamental level, just as the Eleatic Stranger would have it.[31]

28 See *Ep*83.

29 The notion of power (of existing) appears already in one of the four demonstrations Spinoza gives for God's necessary existence (1p11); but given that it is located only in the *second alternative* demonstration, it can hardly be said to occupy a central place at that point in the *Ethics*. For my analysis of power of existing, see below. For an informative discussion on this proof, see Lærke, forthcoming.

30 For an especially noteworthy similar view, see Huenemann 1999, pp. 234, 238. However, Huenemann (pp. 236–7) seems to suggest that *because* God creates by his power everything creatable, each thing is rendered to be necessitated by God's nature; whereas I would like to emphasize that Spinoza's necessitarianism is not subordinate to God's power – identifying God's essence with power is something Spinoza does only *after* having already established that God is a necessary existent from whose nature all possibilities necessarily follow. Most importantly, 1p34's location in the *Ethics* and the way it is proven speaks for this. For a reading that appears to view the relationship between necessity and power as I do, see Scala 1994, p. 28.

31 G. H. R. Parkinson (1981, p. 9) holds that it appears to follow from the triangle analogy of 1p17s that 'Spinoza must speak of the "power" of a triangle, in so far as certain things follow from the essence of a triangle; but we do not normally say that geometrical figures have power.' Without coming to a

We can now see how Spinoza arrives at 1p34: since from God's essence follows the existence both of God (1p11) and of everything possible (1p16), that essence must be causally efficacious (1p34d); thus, God is endowed with power in virtue of his essence, and God's essence and power can be identified. In this way, Spinoza endorses the traditional linkage between power and nature (or essence). However, the distinction between potential and actual does not apply here, just as it does not pertain to the Cartesian idea of force; that Spinoza's God has *potentia* does not mean that he would have a capacity to act or produce an effect which could remain potential, never become exercised. In God-or-Nature, there is no power but *actual* power.[32] Spinoza refers to this when he asserts, 'God's omnipotence has been actual from eternity and will remain actual in the same actuality to eternity' (1p17s). As a consequence, if actualism is defined as the metaphysical position according to which there are no unactualized possibilities, Spinoza can be regarded as an actualist, and also with regard to God's power.[33]

Much of the aforesaid does not pertain to God only: Spinoza thinks that the same essentialist model of power applies to all things. This is witnessed by the general claim that '[n]othing exists from whose nature some effect does not follow' (1p36): when proving this Spinoza builds his case, to an important degree, on the idea that all things 'express' the power of God. The idea behind this will be examined in detail below, but already at this point we can sum up the connection between the interrelated concepts of power, causality, and essence by saying that, in general in the *Ethics*, power means being able to cause effects, and since things cause effects in virtue of their essences, things are intrinsically powerful.[34] Thus, importantly, the linkage

definite conclusion, Parkinson first considers reasons Spinoza might have for saying that a triangle does not have power (perhaps power 'belongs to real things alone, and not to "entities of reason" such as triangles', and the analogy was there 'to illustrate the way in which certain properties follow eternally, and was not meant to be an example of power') and then for saying that a triangle does have power after all. I think Spinoza's position is as follows. The geometrical examples of the *Ethics* are there to explicate the kind of internal structure each thing has, and the necessity involved in the resulting essential causation, but it is important to appreciate the fact that a geometrical object is not a real thing as God is. Power pertains to real things only, certainly not to beings of reason like triangles, and so we should not think that triangles as objects of geometry are endowed with power; thus the first alternative, presented quite perceptively by Parkinson, is the right one. But note that if there existed a (perfect) real triangle, it surely would have power to produce its necessary properties, and were it also *causa sui*, it would have power to exist in exactly the same sense as God does. For his earlier discussion of this topic, see Parkinson 1971, p. 534.

32 Cf. Deleuze 1992 (1968), p. 93; Donagan 1979 (1973), p. 248; Rousset 1994, p. 11; Ramond 2007, p. 147; Di Poppa 2010, p. 277.

33 For a fuller account of this, see Viljanen 2009b.

34 For alternative discussions of Spinoza's concept of power, see Deleuze 1992 (1968), ch. 5; Parkinson 1971, pp. 533–6, 1981, pp. 9–11; Barbone 2002, pp. 102–4; Pietarinen 2003, pp. 143–5; Lin 2004, pp. 29–46.

called into question by Descartes's physics is renewed, namely the one between power and the essence of any finite thing, which makes the route clear for the concept of power to have a proper place and task. However, it is striking – and an indication of the fact that no mere restatement of the old doctrine is here being made – how drastically Spinoza's *a priori* route to the idea that power is something intrinsic in things differs from the scholastic way of starting with sense experience of the activities of ordinary natural things and then concluding that those activities can only be brought about by an internal dynamic principle, the substantial form (this difference, of course, consistently reflects Spinoza's Cartesian approach to epistemological issues).[35] This kind of concept of power belongs not to the realm of physical nature but to metaphysics; however, as I will argue below, Spinoza has good reasons to see it as quite consistent with structural explanations evoked in mechanistic sciences.[36] His overall idea is arguably that when a thing x causes by itself, i.e. in virtue of having the kind of essence or nature it does, an effect E, it is quite natural to say that x has power to E. The interconnectedness of power and essence means that we can say that, for instance, the power of x to E is explained by referring to x's essence.[37]

At this point we may note that Spinoza uses not only 'power' (*potentia*) but sometimes also 'force' (*vis*). There are passages in which he uses the terms interchangeably: 'The force of any passion, *or* affect, can surpass the other actions, *or* power, of a man, so that the affect stubbornly clings to the man' (4p6).[38] Sometimes he simply equates them: '[F]orce, *or* power' (4p60d). '[T]here is also no comparison between the power, *or* forces, of the Mind and those of the Body' (5pr; C, pp. 596–7; G II, p. 280). Thus, it is difficult to say how, exactly, the two notions differ from each other; at any rate, 'power' seems to be Spinoza's choice when he discusses issues pertaining to God and general metaphysics, whereas 'force' sometimes appears in contexts in which the operations of finite things are treated.[39]

[35] See Ch. 1. [36] See Ch. 6.

[37] In putting Spinoza's position in these terms, I have found helpful, first, Kant's (*LM*, pp. 178–9, 183, 328, 376) lectures on metaphysics, in which power is defined neither as a substance nor as a property, but as the internal sufficient ground of a substance to produce accidents as effects, and second, the position of Harré and Madden (1975, pp. 98, 112–13), who themselves (pp. 98–100) note that their theory of causal power and powerful particulars bears a notable resemblance to that of Aquinas. Cf. also: '[I]n some cases at least, what follows from *x*'s nature or essence is in the power of *x*' (Parkinson 1971, p. 534).

[38] See also defaffgen, 4pr (C, pp. 545–6; G II, pp. 208–9), 4p3, 4p5, 4p18d.

[39] It should be noted that apart from *potentia* and *vis*, there is *potestas*, which is also translated as 'power'. As Steven Barbone (2002, pp. 102–4) holds, that notion refers to an individual's control over its environment, obtained by exercising one's *potentia*.

In addition to its already explicated fundamental role, we may observe many other respects in which the notion of power is informative and useful. One is that it brings forward the idea that things are, by nature, genuinely active causers that always can and do make things happen on their own accord – in other words, that reality or nature is intrinsically active. The following passage, while focusing on the admittedly unsurprising denial of the anthropomorphic conception of power, also tells us something crucial about Spinoza's reasons for including the concept in his system:

> By God's power ordinary people understand God's free will and his right over all things which are, things which on that account are commonly considered to be contingent. For they say that God has the power of destroying all things and reducing them to nothing. Further, they very often compare God's power with the power of Kings.
>
> But we have refuted this in 1p32c1 and c2, and we have shown in 1p16 that God acts with the same necessity by which he understands himself, i.e., just as it follows from the necessity of the divine nature (as everyone maintains unanimously) that God understands himself, with the same necessity it also follows that God does infinitely many things in infinitely many modes. And *then we have shown in 1p34 that God's power is nothing except God's active essence.* And *so it is as impossible for us to conceive that God does not act as it is to conceive that he does not exist.*
>
> Again, if it were agreeable to pursue these matters further, I could also show here that that power which ordinary people fictitiously ascribe to God is not only human (which shows that ordinary people conceive God as a man, or as like a man), but also involves lack of power. But I do not wish to speak so often about the same topic. I only ask the reader to reflect repeatedly on what is said concerning this matter in Part 1, from p16 to the end. For no one will be able to perceive rightly the things I maintain unless he takes great care not to confuse God's power with the human power or right of Kings. (2p3s, emphases added)

Thus, the doctrine that the very essence of nature is causal power exercised in a necessary fashion is used to drive home the idea that reality is intrinsically active in character.[40]

In the kind of ontology just outlined, physical and mental phenomena are caused through the action of particulars that possess the power to cause effects according to their essences.[41] I will call this kind of power to cause effects through one's essence *intrinsic power*,[42] and claim Spinoza to

[40] It should be noticed that many French Spinoza scholars have underscored this active nature of substance; see Deleuze 1992 (1968), pp. 90–5, 198; Matheron 1988 (1969), pp. 13–14; Gueroult 1974, pp. 150–1, 188–9.

[41] For similar contemporary ways of characterizing the role and meaning of the concept of power, by which I am influenced here, see Harré and Madden (1975, esp. pp. 112–13), Chalmers (1999, pp. 218–21), and Ellis (2002, esp. pp. 59, 62).

[42] To my knowledge, Juhani Pietarinen (see 2000, p. 67) is the first scholar to have used the term 'intrinsic power' in this sense.

champion *dynamic essentialism.* That the talk of dynamism is here really in order is indicated not only by Spinoza's frequent use of such terms as 'power', 'striving', and 'force', but also by a distinction that can be made between dynamic and pre-dynamic ('non-dynamic' might be even more apposite) thought. After having criticized Locke for thinking physical objects in passive terms, Harré and Madden dub his position as 'pre-dynamic' and write:

> In dynamics the first billiard ball [smacking into the second] can be treated as a genuinely active or powerful thing by virtue of its energy of motion however the motion was acquired. It is a thing that makes things happen, even though what makes it make things happen was not self-initiated or self-created.[43]

By these standards, Spinoza definitely is a dynamist: he claims that even though only God is *causa sui*, once existence is granted to a finite thing, it necessarily causes effects through its essence and is thus, as the analysis of power of acting below will show, genuinely powerful and active.

Another advantage in using the concept of power, especially significant from the third part of the *Ethics* onwards, is that it allows speaking of things and their causality in quantitative terms: it is quite natural to say that things have power to a varying degree (for instance a person, say a philosopher, can be said to have more power of thinking when healthy than when ill); power can, of course, be of a particular intensity.[44] In this respect it differs from such concepts as *essence*, *cause*, and *effect*, which are quite laborious to use quantitatively. In the *Short Treatise*, the expression 'more essence' still belongs to Spinoza's vocabulary,[45] but by the *Ethics* it has vanished; the closest he comes to it is the 'the more reality the essence of the defined thing involves' of 1p16d. That Spinoza also thinks about things in quantitative terms is made clear right from the beginning of the *Ethics*: 1p9 asserts that the more things possess attributes 'the more reality or being' they have, and

[43] Harré and Madden 1975, p. 115. Ellis (2002, pp. 1–3, 59–63) lists Locke among the early modern thinkers who promote what Ellis calls *passivism*, a world-view on which laws of nature are God's commands imposed from the outside on natural things, themselves powerless.

[44] Koistinen (2009b, pp. 157–9) treats forces as intensive magnitudes that, unlike extensive magnitudes, have a certain quantity or intensity but not a part–whole structure (p. 159):

> Force, as well as other intensive quantities, has an interesting ontology. It does not have such a part and whole structure that a force with a certain intensity could be seen to be composed of forces of lesser intensities – i.e., they are not aggregates. However, the ordinary notion of force seems to involve the idea that something can be added to a force and that something can be taken away from it. The intensity of light can become smaller and smaller and can finally be reduced to zero.

In Chapter 6, I will argue that the idea of the *same* power existing with *variable* intensities is focal for Spinoza's theory of individuation.

[45] See e.g. *KV* II.25–6 (C, pp. 145, 147; G I, pp. 107, 110).

the phrase 'the more reality, or being' is repeated in 1p10s. Thus, the being of things admits of degrees,[46] as does their power. Perhaps it could even be said that the being of each thing in endowed with a certain intensity. As an intrinsically causal concept, 'power' is a better choice than 'perfection' or 'reality' when referring to things' causal status; and since that status turns out, in the final analysis, to be the only thing Spinoza regards as ethically significant, it is no wonder that precisely 'power' and related locutions abound in the third and fourth parts of the *Ethics*.[47] As we will see later on, they offer him an appropriate terminology with which to theorize about the morally and psychologically relevant changes we undergo.

Despite the fact that the notion of power appears quite often in Spinoza's writings, and often in important passages, its significance for Spinoza has not always been properly recognized. One reason for this predicament is, I suppose, the fact that from our post-Humean standpoint – which shares with the mechanistic thought the demand to avoid invoking powers when devising explanations for phenomena – it is not particularly easy to say precisely how Spinoza, a rigorous thinker of the mechanistic era, understands the contested notion. In a similar and related vein, even though Spinoza's model of causality differs crucially from the Humean model and is rooted in geometry and mathematical physics, when properly analysed it proves to be much more intelligible than one might expect, and, as I have argued, not without present-day philosophers whose position is in many respects very close to it. In any case, it is the model that underpins Spinoza's thought on power.

To sum up, the concept of power evokes the dynamism pertaining to the essentialist model of causal activity, and it forms an important part of Spinoza's overall project that aims at a sound ontology not afflicted by the obscurities of traditional theocentric (Judaeo-Christian) metaphysics. I hope this section has succeeded in explicating Spinoza's most important reasons for including the notion of power in his metaphysics; however, although this has already demanded unearthing some key parts of the network of concepts in which the notion operates, more analysis is required before its role in Spinoza's thought – especially insofar as it concerns things (modifications) that are limited in character – is sufficiently clarified.

[46] It should be noted that Della Rocca (2008a, p. 49) defends, although by a line of argument different from the one presented here, the idea that, for Spinoza, 'existence itself is not an all-or-nothing affair [...] there are degrees of existence'.

[47] The basic concepts of Spinoza's psychology and ethics are defined in dynamic terms, 'emotion' in 3d3 and 'virtue' in 4d8; see Chs. 5 and 6.

THE POWERFUL EXISTENCE OF SUBSTANCES, MODES, AND ATTRIBUTES

In 1p34d, Spinoza states what follows from the fundamental power of the only substance: all things are and act by the power of God. That existence and acting are fundamental categories pertaining to things is expressed for the first time quite early in the *Ethics*, when Spinoza gives us the definition of freedom:

> That thing is called free which exists from the necessity of its nature alone, and is determined to act by itself alone. But a thing is called necessary, or rather compelled, which is determined by another to exist and to produce an effect in a certain and determinate manner. (1d7)

It is characteristic of Spinoza that he considers power as power to *exist* and as power to *act*; the meaning of this dual character of power will become clear in the course of this study. I begin by examining the linkage between power and existence.

I have argued above that Spinoza thinks that the realization of what is derivable from God-substance's definition requires power; thus all existence has God's power as its basis. This line of reasoning makes its presence felt already in the second alternative demonstration of 1p11, where the concept of power begins its career in the *Ethics*: it is 'known through itself' that '[t]o be able not to exist is to lack power [*impotentia*], and conversely, to be able to exist is power [*potentia*]' (1p11d3). As a consequence,

> the more reality belongs to the nature of a thing, the more powers it has, of itself [*a se*], to exist. Therefore, an absolutely infinite Being, *or* God, has, of himself, an absolutely infinite power of existing. (1p11s)

In a nutshell, power grounds all existence.

As we have seen, the ease with which the notion of power allows describing quantitatively finite things' causal status can be deemed to be one important reason for Spinoza to admit the notion into his vocabulary. In 1p11s he says that the ability of a thing to exist of itself, and hence its power, is equated with the degree of reality of its nature; in other words, *any thing has reality to the extent it has power*. Thus, even though Spinoza does not really rely on the notion of power when proving his necessitarianism, he displays consistent sensitivity to the Eleatic Stranger's test when he construes his theory of real existence, both infinite and finite.

That power is power to bring about existence is not, as such, particularly informative. A more nuanced view on these matters can be obtained through an analysis of the relationship the three basic kinds of entities of Spinoza's ontology – *substances*, *modes*, and *attributes*, of which the first two mentioned qualify as things – have to power. After all, if power brings about, with geometrical necessity, what there is, we should know what it is that is thereby produced. In what follows I aim to show that, strikingly but consistently, all of the three types of entities of Spinoza's system are dynamic in character, each in its own special way.

It can be taken as an interpretative clue that Spinoza does not talk about a thing's power of existing *simpliciter* but equips it with the qualification *a se* ('from itself' or 'of itself') or *in se* ('in itself').[48] There is a piece of philosophical background especially important for discerning the force of that qualification, located in the replies to the first set of objections to Descartes's *Meditations*. When Descartes clarifies his stand on God's existence, he begins by considering whether anything can be the efficient cause of itself. It soon becomes clear that he approaches this issue through an analysis of God's causality and the phrase 'from itself' (*a se*). These are, in turn, to be understood in terms of power. Descartes proclaims that

> there can exist something which possesses such great and inexhaustible power that it never required the assistance of anything else in order to exist in the first place, and does not now require any assistance for its preservation, so that it is, in a sense, its own cause. (CSM II, p. 78)

As he makes clear, this is the positive and philosophically good sense in which existing 'from itself' should be understood. The point Descartes wants to drive home is that all this applies to God, who 'derives his existence from himself, or has no cause apart from himself, depends not on nothing but on the real immensity of his power' (CSM II, p. 80). But in ourselves we do not find power that would preserve us 'even for one moment of time', and so it should be concluded that we derive our 'existence from another being', which, in turn, 'derives its existence from itself' (CSM II, p. 80).

These kinds of considerations are no mere technicalities, and in particular not for Spinoza, as can readily be shown. Now Descartes develops his own version of the ontological framework of substances and accidents – in less technical terms, of things and properties: in the *Principles of Philosophy* (I.51–6; CSM I, pp. 210–12), he mentions properties, qualities, attributes, and modes, but it is uncontroversial that all these refer to what lies on the property side of the traditional substance/accident ontology. God is

[48] Although *a se* occurs in IP11S, Spinoza speaks about being *in se* much more often in the *Ethics* (most notably in ID3 and IA1). I assume the former phrase to refer more to the power of acting, the latter to the power of existing, but, on the whole, Spinoza seems to be using them interchangeably.

substance in the clearest and pre-eminent sense, for he requires nothing apart from himself to exist. Further, and along traditional lines, it is clear that there is a relation of *inherence* between the two kinds of entities: properties, or modes, cannot exist on their own but always have substances as their bearers.

A look at Spinoza's ontology shows it to preserve this ontological architecture practically unaltered. A substance is something which exists in or from itself (1d3); a mode is something which exists in another (*in alio*) (1d5). Now – and as I have argued elsewhere[49] – whereas a mode's way of existing matches the traditional notion of inherence, the substance's way of being (it can be said to *be in itself*, or *exist from itself*) equals what was commonly called *subsistence*. One prevailing understanding of the subsistence/inherence distinction was to see it as a way of distinguishing ontologically independent entities from dependent ones: unlike things that inhere, subsisting entities do not depend on other entities for their existence. Not being dependent on anything else for its existence (save perhaps God) was, of course, considered a traditional mark of substancehood, just as requiring something else in which to exist was a mark of propertyhood.

The interesting point for our purposes is that the basic ontological categories of substance and property (mode), with the corresponding ways of being (subsistence and inherence), have their respective differences with regard to power. It is not especially difficult to see what Spinoza's understanding of existing in itself, or subsistence, amounts to in terms of power: the above mentioned alternative demonstration of God's existence (1p11d3) turns on what 1p11s states in an unequivocal fashion, 'being able to exist is power'. Briefly put, then, *to be in itself is to have power to exist*, or to exist in virtue of one's own power alone. Strikingly, when Spinoza proves the necessary existence of any substance (1p7d) and God (1p11d1), he seems to indicate that it is in some sense impossible to think of a substance as not existing;[50] whatever the full import of this, it can be taken at least to suggest that any substance should be seen as a fundamentally power-laden entity that has causal efficacy to bring about its own existence. To be a substance, to subsist, and to be endowed with power to produce existence all fall into one.

Consistently (and understandably) enough, only God, the sole thing that can be said to subsist – to be purely in itself – in Spinoza's system, is endowed with this kind of power.[51] God-substance is an inexhaustible

[49] Viljanen 2009c; see also Jarrett 1977; Carriero 1995.

[50] In many discussions, Koistinen has drawn my attention to this.

[51] It should be noted that Spinoza's argument for monism in 1p14 is not based on power; however, his answer to an often presented objection to the argument might well be (see Garrett 1979, pp. 217–20). For a reading that emphasizes the linkage between divine self-causation and Spinoza's theory of power, see Lærke, forthcoming.

source of causal power to produce existence, capable of realizing not only himself (subsistence) but an infinite number of finite things (that inhere in God) as dictated by his nature. Given that Spinoza equates God's power with his essence (1p34), and given the close linkage between a thing and its essence (there is, at most, something akin to the distinction of reason between the two), it follows that *Spinoza's God is, in essence, a power – the ultimate dynamic factor behind all existence*. This may sound radical, but as we saw, Descartes also understands only God to be capable of existing from himself, because only his existence depends on 'the real immensity of his power' (CSM II, p. 80) alone. Spinoza thus simply presents his own (immanentist and geometry-inspired) variant of a time-honoured thesis accepted not only by Descartes but more generally by the Neoplatonic and scholastic traditions of philosophical theology:[52] only God exists necessarily, in virtue of his own power alone.

The case of finite modes – which exist in another – is far more complex. In the Aristotelian-scholastic tradition, that accidents inhere in a substance certainly does not entail that they are deprived of power; similarly, the fact that Spinozistic modes inhere in God does not make them powerless. This power, however, has its limits, for finite things are such that they 'come to be from external causes' and thus

> owe all the perfection or reality they have to the power of the external cause; and therefore their existence arises only from the perfection of their external cause, and not from their own perfection. (1p11s)

In other words, finite things' power *has nothing whatsoever to do with the beginning of their temporal existence*; their actualization can be brought about solely by the efficacy and power of external causes, the chain of which continues without end (1p28). To a certain extent, Spinoza thinks along the same lines as Descartes for whom finite things are not causes of their own existence; the cause must be another being, ultimately one that exists from itself, namely God (CSM II, pp. 79–80).[53] So, for Spinoza, finite things – or their essences – are not causal factors in bringing about the beginning of their existence. However, when we turn to consider issues pertaining to *persisting* in that existence, things become considerably more complicated, and fascinating.

[52] See Kristeller 1984, p. 7; Carriero 1991, p. 56, 1994, pp. 626, 639. This does not mean that there would not have been significant differences in the precise way God's causality was conceived; see Lærke 2009, esp. pp. 174–9.

[53] See also CSM II, pp. 118, 165.

I will discuss the famous *conatus* proposition in detail from Chapter 4 onwards, but here it should be noted that the proposition contains a noteworthy reference to a notion that has served as the guiding thread for this section, namely *being in itself*: 'Each thing, *insofar as it is in itself* [*quantum in se est*], strives to persevere in its being' (3p6, translation modified, emphasis added).[54] But here we encounter a problem: how can modes be in themselves to *any* degree, given that they are defined as beings that are *in another* (1d5)? Certainly, we have seen that no mode can be in itself the same way as God is in himself. Here it all boils down to the meaning of the mitigated form of being *in se*, as expressed in 3p6. But how should we understand it? Nothing in the definitions presented in the beginning of the *Ethics* implies that the distinction between being in itself and being in another would be anything but a sharp dichotomy, allowing no degrees. Finding an answer to this question reveals the sense in which finite modes have power to exist.[55]

As I. Bernard Cohen explains, the expression *quantum in se est*, adopted by Descartes (and later by Newton), originates from Lucretius and

> had been conceived in the traditional physics as 'natural', or 'according to its nature', or 'by nature', or 'by its own force', or 'by itself' – 'without external force'.[56]

The seventeenth-century understanding of the phrase remained largely unchanged, so that in many contexts it 'meant "naturally" or "by nature", or "without external force"';[57] further, Cohen also suggests that *quantum in se est* equals '"so far as it rests with themselves" as opposed to something external'.[58] All this, together with the way in which Spinoza argues for the *conatus* principle in the demonstration of 3p6,[59] strongly suggests that Spinoza's understanding of the meaning of *quantum in se est* diverges little from the tradition and that he means by the phrase, roughly, 'by nature', 'insofar as a thing's nature is considered' or even 'to the extent a thing

[54] Curley's translation reads: 'Each thing, as far as it can by its own power, strives to persevere in its being' (C, p. 498). For reasons that will shortly become clear, I do not find this an inappropriate way to translate the proposition. However, for purposes of analysis, I think it is helpful to opt for the more technical expression; cf. Della Rocca 2008b, p. 323.

[55] For discussions on issues pertaining to these matters, see Garrett 2002, pp. 134–41, 144; Lin 2004, pp. 39–40. In his important paper, Garrett (2002, p. 144) acknowledges the distinction between coming to existence and continuing in existence, but he does not further analyse this distinction and its relation to his interpretation of Spinoza's finite things as 'quasi-substances'.

[56] Cohen 1964, p. 147. As Cohen notes, in Aristotelian physics this referred to the contrast between 'natural' and 'violent' motion.

[57] Ibid.

[58] Ibid., p. 144. Cohen's suggestion is based on a note by Cyril Bailey, a translator of Lucretius.

[59] For a detailed discussion of the argument, see the next chapter.

operates in virtue of its nature alone, with no regard to external causes'. So although finite things are not substances but modifications of substance in which they inhere, they still have their own essences that are causally efficacious. We can now see what a finite thing's essential causal power can amount to with regard to existence: were any actualized finite thing to exist unaffected by external causes, by reason of its own nature alone, it could never be destroyed. In the following passage Spinoza speaks of human beings, but the point can be generalized to all finite individuals:

[I]f it were possible that a man could undergo no changes except those which can be understood through the man's nature alone, it would follow (by 3p4 and p6) that he could not perish, but that necessarily he would always exist. (4p4d)[60]

As it is true that finite things cannot begin their actual existence by their own nature, so it is also true that their existence cannot come to end by it; in the absence of external causes, an actual finite thing exists without end, or sempiternally. The important idea is that although no finite thing can *produce* its actual existence – in fact, to do that its essence would have to involve existence, which would make it an *ens necessarium* – it can, however, *persist* in existence in virtue of its essence, due to the causal efficacy pertaining to that essence. It can thus be said that in a sense finite things are not in and from themselves, for their actual existence begins by reason of external causes; but in another sense they are in and from themselves, for if not entirely overcome by external causes, they continue to exist in actuality by reason of their own essence, because that essence is a power. In brief, that finite modes have power to exist means that they are endowed with power to continue their actual existence.

Now finite things are modes or affections of the infinite substance, their power inseparable from God's power; for instance a human being's power 'insofar as it is explained through his actual essence, is part of God *or* Nature's infinite power' (4p4d).[61] But how exactly do finite things modify, or partake in, God's power? As highlighted here, Spinoza puts much weight on the idea that essences or natures are the loci of causal power. Given the tight connection between essences and powers, Spinoza's position may be labelled *dynamic essentialism*. But how do essences acquire their power? Recall Spinoza's definition of essence:

I say that to the essence of any thing belongs that which, being given, the thing is [NS: also] necessarily posited and which, being taken away, the thing is necessarily

60 See also *TdIE* § 57.

61 As Garrett (2002, p. 140) states, 'whatever power singular things have is at the same time *also* (a share of) God's power, power that God expresses *through* singular things'.

[NS: also] taken away; or that without which the thing can neither be nor be conceived, and which can neither be nor be conceived without the thing. (2d2)

This passage leaves much room for interpretation. According to Bennett, the essence of a singular thing (as stated in 2d2) is '[t]hat set of its properties which are jointly sufficient and severally necessary for its identity'.[62] However, keeping the linkage between power and essence in mind, it is highly unlikely that Spinoza is here talking about qualities or properties static by their nature but about something much more dynamic. Accordingly, 2d2 can be interpreted as saying that any thing is necessarily posited by its intrinsic power and necessarily taken away when this power is removed; nor can God be, without the thing, in that 'certain and determinate' way he is when the thing begins its existence. I would argue that the idea here is that the essences of finite things indicate the *manner* in which intrinsically active reality must be under a certain attribute for specific finite things to exist. This is to take Spinoza quite literally, for singular things are, for him, ways in which God is modified.[63]

The following line of argument, extractable from the opening part of the *Ethics*, can be seen as lurking behind 2d2. Everything there is follows from the essence of God, which means that God is intrinsically powerful (1p16, 1p34), and since finite things are expressions of this infinite power of God (1p25c, 1p36d), they can quite plausibly be described as specifically modified portions of the total power of nature.[64] This means that in the Spinozistic scheme of things, finite individuals can be conceived as specifically determined centres of causal activity and power, individual essences operating as modifiers determining the way in which substance and its efficacy or total power is distributed.[65] Understood in this way, a finite thing cannot be distinguished from a specifically modified portion of God's power, or a certain manner of operation and causation; as we have seen, already persisting in 'mere existence' requires constant causal activity, or a certain kind of power to exist.

62 Bennett 1984, p. 233. Cf. Garrett 2002, p. 130.

63 Cf. 1p28d that says finite things to follow from 'God or an attribute of God insofar as it is *modified by a modification* which is finite and has a determinate existence' (emphasis added).

64 Della Rocca (2003a, p. 79) writes similarly, 'finite objects can [. . .] be causes simply because their power is, as it were, a chunk or aspect of God's power'.

65 Pietarinen has stressed to me that this approach opens up promising possibilities for interpreting the famous and difficult 2le7s, in which Spinoza says the whole of nature to be 'one Individual, whose parts, i.e., all bodies, vary in infinite ways, without any change of the whole Individual'. On the present interpretation, this may be taken to say that the amount of nature's infinite power does not alter, but it is *distributed* or *ordered through* actual essences, the frame of distribution (i.e. the order of actual essences at a given time) changing from one moment of time to another.

I believe that Spinoza thinks that it is possible to conceive any finite modification in a reified way, as a certain thing, or in a dynamic way, as a centre of causal activity, and that he regards the latter perspective as the metaphysically more adequate one, revealing something fundamental about the inner workings of things.[66] Recall the familiar proposition: 'Nothing exists from whose nature some effect does not follow' (1p36), because 'whatever exists expresses in a certain and determinate way the power of God' (1p36d). A passage located in a much later demonstration sums it all up nicely: human beings' power is really God's power, 'not insofar as it is infinite, but insofar as it can be explained through the man's actual essence' (4p4d).[67] So, things are essential causers and effects follow from finite essences, because *essences individuate things by indicating how God-or-Nature's power is modified.*[68] This is what causation 'through essence' means.

We saw the gist of the linkage between Spinoza and certain scholastics to lie in the fact that Spinoza's talk of 'properties following from essence' corresponds to the medieval doctrine of essential principles and properties flowing from them. Now we can see that Spinoza's unwillingness to think of essences as sets of properties stems from the idea that all things are modifications of power, and hence certain effects follow from their essential, organized activity. In an important passage, Charles Jarrett claims:

> Spinoza's conception of what pertains to the essence of a thing is simply not the same as our (allegedly muddled) notion of a 'de re' necessary property. That this is so should be clear from the claim in 2p37 that things (properties) that are common to everything constitute the essence of no singular thing. Thus, for example, extension (which constitutes the essence of God) is essential to every body, but it does not constitute the essence of any finite body. What constitutes the essence of a finite object is thus not just any property that the object necessarily has. It is rather something unique to the individual – something which, when given, the thing is given, and something from which every property of the object (*when considered alone*) can be concluded.[69]

This is right, and I would only add that the 'something unique to the individual' is a specific modifier of God's power that posits a thing as a causally efficacious centre of operation.

[66] Cf. Barbone 2002, pp. 96, 100; Koistinen 2009a, p. 185; Lærke 2009, pp. 180–1, 188–9.

[67] Cf. Forsyth 1974 (1948), p. 8; Deleuze 1992 (1968), p. 227; Moreau 1975, p. 74; Rousset 1994, pp. 15, 21.

[68] Following Hector-Neri Castañeda (1975, pp. 132–3), I understand the problem of individuation to concern describing that which *makes* an individual individual, i.e. what constitutes a thing's individuality; this problem should be kept separate from the problem of how two distinct individuals are *differentiated* from each other. Clearly, essences are, for Spinoza, the individuating ingredient in things.

[69] Jarrett 2009, pp. 132–3.

If the connection between essence and power is as close as argued here, what should we say about the contention, obviously a fundamental one in Spinoza's metaphysics, that attributes constitute the essence of God (1d4, 1d6)? Given dynamic essentialism, we should expect Spinoza to view also attributes as somehow powerful; and indeed, we find him writing in his correspondence as follows:

[F]rom Extension as conceived by Descartes, to wit, an inert mass, it is not only difficult [. . .] but quite impossible to demonstrate the existence of bodies. For matter at rest, as far as in it lies, will continue to be at rest, and will not be set in motion except by a more powerful external cause. For this reason I have not hesitated on a previous occasion to affirm that Descartes' principles of natural things are of no service, not to say quite wrong. (*Ep*81; S, p. 956)

Still more important is the highly exciting – and difficult – passage included in the next letter he sent to Tschirnhaus:

With regard to your question as to whether the variety of things can be demonstrated a priori solely from the conception of Extension, I think I have already made it quite clear that this is impossible. That is why Descartes is wrong in defining matter through Extension; *it must necessarily be explicated through an attribute which expresses eternal and infinite essence.* (*Ep*83; S, p. 958, emphasis added)

I think these passages strongly point to the direction of my interpretation. They claim, firstly, that extension should not be thought of as inert mass (*Ep*81), and secondly, that matter cannot be derived from the concept of extension alone for it follows from an attribute *which expresses God's essence* (*Ep*83) – that is, given 1p34, God's power. Spinoza seems to imply that something more fundamental than matter or extension is involved here, and my contention is that he is pointing precisely to the nature of substance as active power; this is what Spinoza means when he speaks (in his idiom) of matter being 'explicated through an attribute which expresses eternal and infinite essence' (*Ep*83; S, p. 958). I would thus agree with Huenemann who comments here, '[t]his eternal and infinite essence is presumably God's essence, which is the same as his power'.[70] Spinoza's point is that since extension, as an attribute, expresses God's essence or power, matter should be understood through a dynamic conception of extension, that is, through extension as an active attribute that needs nothing external to itself to generate the variety of corporeal things.[71]

[70] Huenemann 1999, p. 234.

[71] This line of thought has had its proponents on both sides of the Atlantic; see Wolf 1974 (1927), pp. 21–4; Matheron 1988 (1969), p. 13; Harris 1972, p. 195; Gueroult 1974, pp. 150–1, 188–9; Lennox 1976, p. 492; Della Rocca 2003b, pp. 225–6; Deveaux 2003, p. 334; Rizk 2006, p. 44; Marshall 2008, p. 70; Di Poppa 2010, pp. 281–3.

POWER OF ACTING

Apart from power to exist, Spinoza speaks about power by which things act or operate. Although he often speaks separately about power of existing and power of acting, the power by which God exists is also the power by which he acts:

> That eternal and infinite being we call God, *or* Nature, acts from the same necessity from which he exists. For we have shown (1p16) that the necessity of nature from which he acts is the same as that from which he exists. The reason, therefore, *or* cause, why God, *or* Nature, acts, and the reason why he exists, are one and the same. (4pr; C, p. 544; G II, p. 206)

So, ultimately God's power to exist and to act amount to one and same thing: God exists absolutely and is fully active, and there is only one unified power of God. However, with regard to finite things we must take into account that, as already noted, they can be self-determined or 'in themselves' to a varying degree; their level of activity can be, and is, determined in different ways by other finite things. Here the concept of power of acting (*potentia agendi*) comes to the fore: after the second part of the *Ethics* Spinoza quite often refers to power in the sense of power of acting; as this notion is absolutely central for his moral psychology, we should know what Spinoza means by it.

There is, of course, the traditional scheme operating with the distinction of activity and passivity, and since *acting* in *power of acting* obviously refers to activity, we should obtain at least a rough working understanding of Spinoza's view on the time-honoured distinction, and of the way in which Spinoza's stand relates to that of his predecessors. Spinoza defines action and passion as follows:

> I say that we act [*agere*] when something happens, in us or outside us, of which we are the adequate cause, i.e. (by d1), when something in us or outside us follows from our nature, which can be clearly and distinctly understood through it alone. On the other hand, I say that we are acted on [*pati*] when something happens in us, or something follows from our nature, of which we are only a partial cause. (3d2)

This means that x is active when an effect E is brought about by its essence alone; when this happens, x is the adequate, i.e. complete and sole, cause of E. And x is passive when it brings about together with external causes a state, or E, in itself. So, both activity and passivity are defined by referring to natures or essences. Here, once again, the essentialist model of causation presented above makes its presence felt, as the expression 'when something [. . .] follows from our nature', mentioned twice in the definition, makes clear. We act in the same sense as God does when an effect follows from our nature alone, that is, when we have power to bring about that effect with no contribution from other causes. In the end, this is what acting means. The fundamental case of this kind activity is, of course, the pure action of the only substance by which infinitely many effects, i.e. being with a strictly determined structure, is brought about. While 3d2 reveals that patients contribute to causal processes as well – they are thus by no means causally inefficacious – Spinoza still thinks it makes sense to talk about the active and passive aspects of a causal phenomenon.

I would suggest that Spinoza's view of the active–passive distinction can be expressed roughly as follows: *the active aspect of a causal occurrence is the unaffected exercise of intrinsic power; if there is a patient present,*[72] *only its state changes, although the exact character of the resulting state also depends on what kind of thing the patient is and what kind of causal resources it has.* Nothing here seems to go against either Descartes's or Hobbes's views[73] – or even against the standard Aristotelian-scholastic view, which may be formulated as follows: 'Typically, substances (agents) act upon other substances (patients) to bring about or actualize or produce states of affairs (effects). So both agents and patients may properly be said to contribute causally to the effects produced.'[74]

Before moving on, we should recall a central feature of the essentialist model of causation on which Spinoza's theory of power of acting is, to an

[72] Note that when we, by ourselves, make something happen in us, no patient is required.

[73] There is, however, the notable difference that whereas Spinoza begins his definition by indicating that we may (actively) cause something in us (in other words, that also finite things can be immanent causes), Descartes and Hobbes seem to have only transeunt causation (i.e. causation taking place between two distinct substances or things) in mind when they discuss activity and passivity. Especially Hobbes gives this impression:

> A BODY is said to work upon or *act*, that is to say, *do* something to another body, when it either generates or destroys some accident in it: and the body in which an accident is generated or destroyed is said to *suffer*, that is, to have something *done* to it by another body; as when one body by putting forwards another body generates motion in it, it is called the AGENT; and the body in which motion is so generated, is called the PATIENT; so fire that warms the hand is the *agent*, and the hand, which is warmed, is the *patient*. That accident, which is generated in the patient, is called the EFFECT. (*De Corpore* II.9.1; *EW* I, p. 120)

And slightly later he contends, 'a CAUSE simply, or *an entire cause, is the aggregate of all the accidents both of the agents how many soever they be, and of the patient, put together*' (II.9.3; *EW* I, p. 120). Descartes, in turn, writes in *The Passions of the Soul* (§ 1):

> I note that whatever takes place or occurs is generally called by philosophers a 'passion' with regard to the subject to which it happens and an 'action' with regard to that which makes it happen. Thus, although an agent and patient are often quite different, an action and passion must always be a single thing which has these two names on account of the two different subjects to which it may be related. (CSM I, p. 328)

[74] Freddoso 1988, p. 79.

important degree, based. Spinoza's model, harking back to the scholastic doctrine of emanation, brings forward the idea that things are active by their nature: in virtue of their essences alone, things are necessarily prompted to do what they are able to. This line of thought goes strongly against the familiar Thomist tenet, 'what is in potency cannot reduce itself to act' (*PN* III.18; *SW*, p. 14). This is a principle that Spinoza could not accept, even if it were purged of teleology: his contention is precisely that *nothing extrinsic is needed for a thing to act.* For Aristotle as for Aquinas there must be something actual accounting for the beginning of the actualization of a natural agent's potentiality;[75] '[t]he Aristotelian does not suppose that the inceptions of a thing's changes can be referred to that thing's nature alone'.[76] Of the early modern thinkers, at least Hobbes holds similarly in *Of Liberty and Necessity*, 'I conceive that nothing taketh beginning from *itself*, but from the *action* of some other immediate *agent* without itself' (*EW* IV, p. 274). In Spinoza's view, this kind of being set underway by something extrinsic is not necessary to attain activity, for we are active simply to the extent that we are left to our own devices. Hence, a state of potentiality and passivity does not necessarily precede activity nor is anything but the thing itself needed for 'getting it going'. These contentions are in tune with the traditional doctrines of emanation.

It is not particularly difficult to acquire a rudimentary understanding of the notion of power of acting. Since 'power' means, roughly, being able to cause effects, and 'acting' means being the sole cause of an effect by one's nature, it follows that having power of acting means being the complete cause of an effect in virtue of one's essence alone, while the specific nature of that effect depends on what kind of essence one has. This might also be called the *agent's power* to bring about effects of its own accord, with no regard to other things. All this, of course, is completely in line with, indeed a facet of, Spinoza's dynamic essentialism. Now, it is fairly easy to see what this kind of agent's power is in the case of the only substance, i.e. God-or-Nature. As a purely

[75] See the beginning of this chapter. Here applies the so-called principle of prior actuality, which can be formulated as 'what is potentially F becomes actually F by the agency of something that already is actually F' (Emilsson 2009, p. 82). Susan James (1997, pp. 30–7) offers a very instructive discussion on activity and passivity in Aristotle; as she notes (p. 35), Aristotle does not think that our cognitive powers would require something external to make them active: human knowledge can be exercised at will, provided that there are no impeding factors.

[76] Des Chene 1996, p. 22. However, as Des Chene points out, this does not apply to living things; they are capable of self-movement, which was regarded as 'the delimiting feature of animate things' (p. 22). Spinoza's 'activity thesis' thus squares well with his famous contention that all individuals are at least to some degree animate (2p13s), although his argument for the latter contention, as I understand it, takes another route: because God has an idea of each and every thing, there is no thing without its idea or 'soul'.

active thing, God's power of acting is absolute, which means being the adequate cause of everything that exists through the divine essence alone.

All finite things, in turn, are effects of God's essence, expressing his power, and therefore ontologically speaking not substances but modifications or properties of substance. The real challenge is to provide a convincing account of finite things' power of acting, for it, not God's power of acting, is of crucial importance for Spinoza's ethical theory. However, this is no simple task for the reason that although God and human beings as God's modifications all have essences from which 'certain things necessarily follow' (3p7d), God is causally isolated and endowed with unchanging and infinite power of acting, whereas we are in interaction with other finite things, which has an effect on the way in which our power is determined. In outlining such an account, it is necessary to refer to passages where Spinoza deals primarily with human beings' power of acting, but his points are in many places quite general and apply to all finite things.

That Spinoza equates human power of acting and virtue is made clear more than once in the *Ethics* (3p55s, 3p55cd, 4p52d), and virtue, for its part, is defined as follows:

> By virtue and power I understand the same thing, i.e. (by 3p7), virtue, insofar as it is related to man, is the very essence, *or* nature, of man, insofar as he has the power of bringing about certain things, which can be understood through the laws of his nature alone. (4d8)

In accordance with this, in the demonstration of 3p54 Spinoza identifies the mind's power of acting with what the mind can in the strict sense do. So, as should be expected, we, too, have power of acting, or power to cause effects that are (in Spinoza's idiom) understood through the laws of our nature alone. As the reference to laws suggests, having this kind of power amounts to causing effects *autonomously*.[77]

As noted above, insofar as we have power of acting, we are active. Hence, increases and decreases in power of acting correspond to increases and decreases in the degree of our activity (and passivity). This must, I think, be conceived as follows. On the one hand, when x's power of acting is great, x is to a large extent active and mostly brings about effects that are determined by x's own nature alone. I would suggest that this kind of high level of activity equals exercising one's power mainly either in being a complete cause of one's *own* states (this kind of immanent causation takes

[77] As Matheron (1988 [1969], p. 50, translation mine) rightly observes, power of acting equals 'the individual's aptitude to do what follows from the laws of its nature alone'. As Matheron notes, this equals autonomy of behaviour.

place, for example, when we form clear and distinct ideas of geometrical objects in our minds) or in adequately causing certain aspects of effects that inhere in other things (this kind of transeunt causation takes place, for example, when a constructor builds a house to live in). On the other hand, when x's power of acting is low, x is to a large extent passive, and its states are mainly joint products of x's nature and of the natures of other things (for example, when someone is overwhelmed by the power of harmful external causes to the extent that a state of depression, or melancholy, develops). This implies that from x's essence effects always follow, but insofar as x is under the influence of external causes, these effects are interactive mixtures of x's causal power and external causes' power, and passions of x.[78] At the limit, x's power of acting can be diminished to zero, but this does not necessarily equal x's destruction, only that no effect is produced by x's power *alone*. Death (i.e. the end of actual existence) results when x is no longer capable of *opposing* – whether this occurs without or, as it practically always does, with the help of useful external causes – those harmful external causes that are disposed to bring about the destruction of x. In other words, even when x's power of *acting* is reduced to nothing, it does not follow that x would not have power to be a causal factor in producing effects that are its passions.[79]

Spinoza discards the potential–actual framework, which, as we saw, has an effect on how he conceives God's power: it is purely actual, which equals realizing everything possible. Similarly, finite things' power cannot be to any degree potential: *all of our power is always fully exercised.* But this intrinsic power can become determined, and thus exercised, in different ways.[80] The unaffected part of any individual's, say x's, intrinsic power can only produce actions, the nature of which depends wholly on x's essence. When y determines x's power and thus causes a passion in x, this affection usually leaves more or less of x's intrinsic power untouched. Any passion

[78] Della Rocca (1996, p. 211) holds, quite plausibly I think, that a causal factor can be an active cause of a certain effect to a varying degree:

> [W]e can define an increase in *power* of acting in the following way: 'An object comes to have a greater power of acting to the extent to which it comes to be able to be active to a greater degree with regard to a certain effect.' In other words, something's power of acting increases to the extent to which it becomes less dependent on external things in the production of some effect.

Cf. also: 'Power of acting can thus be understood as admitting of degrees. The more the power or essence of a thing explains its effects, the more power of acting it possesses. The more external causes help determine its effects, the less power of acting it has' (Lin 2006, p. 326).

[79] For more on this and related issues, see Ch. 6. Cf. also Della Rocca 1996, p. 212.

[80] I believe my account is in accordance with those of Deleuze (1992 [1968], esp. p. 93) and Donagan (1979 [1973], pp. 248–9).

expresses the power of y in x; a positive passion increases x's amount of actively exercised intrinsic power, and a negative passion indicates to what extent y is able to prevent x from using its intrinsic power in a self-determined way. The power of acting expresses the activity of the individual, whereas passions, expressing the power of external causes, indicate what the individual is not determined to do by its own power alone. In the end, intrinsic power and its relation to the power of external causes forms the basis of finite existence.

In this chapter I have argued that in Spinoza's dynamic essentialism, individuals are fundamentally beings of power that have no choice but to operate according to their natures. In this important respect, then, Spinoza is much closer to Leibniz than to Hobbes. Although our intrinsic power cannot be the cause of the beginning of our actual existence, as long as our essence is instantiated in actuality, it is manifested as power to exist, or as the ability to cause the continuation of our actual existence. Moreover, insofar as we are unaffected by external causes, our power is that of acting. However, because we are finite entities our power cannot escape being determined by innumerable external causes.

What I take to be established at this point is that the concept of power steps in at a very basic level of Spinoza's system: power is required to realize God and his modifications. Thereby is formed a system of being akin to the one internal to geometrical objects, and one that is the source of intelligibility, necessity, and order in reality. Indeed, this is the first plank of Spinoza's geometry of power. It is particularly relevant for the remaining chapters of this work that we have seen Spinoza derive from his basic ontological commitments a specific conception of essences and powers, because it forms the conceptual machinery constantly at work in his theory of temporal human existence. There our determinations amount to change which pertains properly to things considered *sub specie durationis* only. There are times when we have a great deal of power of acting, i.e. an agent's capacity to bring about effects the nature of which depends on our essence alone; but more often at least a considerable amount of our power is engaged in bringing about passions, positive or negative, that are joint products of our and other things' intrinsic powers. As we will see, the overarching idea of Spinoza's ethical project can be said to be to achieve a maximally high level of power of acting. But developing a fully fledged theory of human existence requires considerable elaboration, and even reconsideration, of the power involved, and its relation to the geometrically modelled being of things. When what is decreed from eternity is played out in the temporal register, it is *power as striving* that carries the major metaphysical load.

CHAPTER 4

The derivation of the conatus *doctrine*

There can be no doubt that the concept of striving – *conatus* – is central for Spinoza's project: in brief, his psychology and ethics are based on it. Thus it is no wonder that *conatus*, and via it the concept of power, have increasingly often figured in Spinoza scholarship, on both sides of the Atlantic.[1] My main contribution to this discussion rests on the contention that an adequate understanding of the *conatus* principle requires seeing it as something that emerges from certain key aspects of Spinoza's ontology with which we are now familiar. As a matter of fact, *conatus* is the most prominent manifestation of what I have called Spinoza's geometry of power, for it is the feature through which Spinoza's rationalist ontology is applied to temporal existence. Of course, discerning how exactly this happens is no small task, but I believe it is well worthwhile, for the result is a unique and truly compelling view of finite individuality and human agency.

As already noted, the main proposition of the *conatus* doctrine states: 'Each thing, insofar as it is in itself, strives to persevere in its being'[2] (3p6, translation modified). This striving, in turn, is designated as the *actual essence* of all things, human beings included (3p7);[3] in other words, actual Spinozistic things are all *strivers* of different kinds. However, Spinoza's argument for 3p6 has been severely criticized for being defective in many ways. The relevance of all the elements it consists of is by no means evident,

[1] For readings in which the concepts of power or striving play a prominent role, see Wolf 1974 (1927); Deleuze 1992 (1968) and 1988 (1981); Matheron 1988 (1969), 1991a, and 1991b; Parkinson 1971 and 1981; Negri 1991 (1981); Bove 1996; Della Rocca 1996 and 2003b; Huenemann 1999; Ramond 1999; Rice 1999; Barbone 2002; Deveaux 2003; Pietarinen 2003; Youpa 2003; Jaquet 2004 and 2005; Lin 2004 and 2006; Sévérac 2005; Spindler 2005; Rizk 2006; Garrett 2008; Koistinen 2009a and 2009b; Lærke 2009 and forthcoming; LeBuffe 2009 and 2010.

[2] '*Unaquaeque res, quantum in se est, in suo esse perseverare conatur*' (G II, p. 146).

[3] This idea is expressed already in the *Metaphysical Thoughts* as follows: 'For though the thing and its striving to preserve its being are distinguished by reason, or rather verbally (which deceives these people very greatly), they are not in any way really distinct' (*CM* I.6; C, p. 314; G I, p. 248).

and scholars disagree as to which ones are truly important for the main proposition. Also the consistency of Spinoza's derivation has been questioned. For Spinoza, these accusations are not a minor problem: as it is a central undertaking of his to derive, in geometrical fashion, true ethics from sound metaphysical principles, what is at stake here is nothing less than the overall cogency of his system. Thus, were Spinoza's arguably most ardent contemporary critic, Bennett, right in claiming that the derivation of the *conatus* doctrine is 'irreparably faulty', it would also be justified to hold, as Bennett does, that '[t]he *Ethics* in fact is broken-backed'.[4] Against this, I will argue that having a firm grasp of Spinoza's concept of power enables us to discern the argument's general idea. Further, it seems to me that thus far the structure of the derivation has not been completely correctly understood, so showing how its various ingredients work together is an important task I will undertake. Before all this, however, some general remarks on the doctrine and its historical sources are in order.

STRIVING IN THE PHILOSOPHICAL TRADITION AND IN SPINOZA'S SYSTEM

That some kind of striving pertains to all or at least to very many things is far from being Spinoza's invention; it has even been claimed that '[a]t the time of Spinoza the principle of self-preservation became a commonplace of popular wisdom'.[5] The predecessors of Spinoza's doctrine can be schematically divided into pre-modern and modern ones, and this line of division has not only to do with chronology but also with how things' striving was conceived.

The idea that every natural thing has its characteristic impulse or inclination to act was accepted by the Peripatetics. Aquinas's position exemplifies this well:

> It is necessary to assign an appetitive power to the soul. To make this evident, we must observe that *some inclination follows every form*: for example, fire, by its form, is inclined to rise, and to generate its like. Now, the form is found to have a more perfect existence in those things which participate [in] knowledge than in those which lack knowledge. For in those which lack knowledge, the form is found to determine each thing only to its own being – that is, to its nature. Therefore this *natural form is followed by a natural inclination, which is called the natural appetite.*

[4] Bennett 2001, p. 222. In fact, Bennett (1984, pp. 234–46) thinks that 3p6 and its demonstration contain altogether four fallacies of equivocation; and as Della Rocca (1996, pp. 200–6) has identified still one more apparent equivocation in it, Garrett (2002, p. 128) is right in concluding, '[t]he argument thus appears to be one of the most egregiously equivocal in all of early modern philosophy'.

[5] Wolfson 1961 (1934) II, p. 196. Cf. also: 'As we have seen, Spinoza is rather casual about proving the *conatus* doctrine. One reason for that may be that he found versions of the principle widely accepted by previous philosophers' (Curley 1988, p. 113).

But in those things which have knowledge, each one is determined to its own natural being by its natural form, in such a manner that it is nevertheless receptive of the species of other things. (*ST* I, Q. 80, A. 1, emphases added)

However, although, as Wolfson observes, not only Aquinas but also Augustine and Duns Scotus see a connection between appetite and self-preservation,[6] the Stoic influences on Spinoza's doctrine are generally the pre-modern ones emphasized.[7] This is understandable for the simple reason that Cicero uses the same specific Latin term as Spinoza does some 1,700 years later. The important passage explains,

just as the several departments of nature are sprung from their own seeds, and thrive within the limits prescribed by them, so the nature of the universe as a whole has its own chosen movements, those strivings [*conatusque*] and desires which the Greeks call *hormai*, and it directs its actions in accordance with these, just as we ourselves do when we are stirred by our spirits and feelings. (*NG* II.58, translation modified)[8]

Cicero also writes about animals having a natural appetite to strive for suitable food (*NG* II.122) and reports that '[e]very animal, as soon as it is born (this is where one should start), is concerned with itself, and takes care to preserve itself' (*ME* III.16).[9] Moreover, appetite is, along with reason, an essential mental force snatching 'a man this way and that' (*Duties* I.101). So clearly, striving for self-preservation is, for the Stoics, something that living things are prompted to do by their nature, and a certain mental power is its important manifestation.

Already these quick observations show that philosophers of Antiquity and the Middle Ages provide an abundance of material for the *conatus* doctrine; often, the basic striving of natural things was called 'inclination' or 'appetite'. Keeping the penetrating influence of Aristotelianism and Spinoza's self-confessed knowledge of Stoic philosophy in mind,[10] it would be astonishing had he not been well aware of the pre-modern thought related to his own doctrine. However, as also the just cited passages testify, the Stoics talk about living (or ensouled) beings as having *conatus*, whereas for Spinoza the

[6] Wolfson 1961 (1934) II, p. 196. See also Des Chene 1996, pp. 235, 239; Pietarinen 2003, pp. 137–8; Lin 2004, pp. 46–9.

[7] For discussion on Spinoza's *conatus* doctrine and Stoicism, see Wolfson 1961 (1934) II, pp. 196–200; Kristeller 1984, p. 5; Curley 1988, pp. 114–15; James 1993, pp. 296–7; Long 2003, p. 374; Nadler 2006, p. 196. For an instructive recent comparison of Spinoza's monism and that of the Stoics, see Miller 2009.

[8] Wolfson (1961 [1934] II, p. 196) refers to this passage when he claims that *conatus* (and *appetitio*) are Cicero's translations of the Greek *hormê*. See also Oittinen 1994, p. 180.

[9] See also *ME* III.20, IV.16.

[10] See 5pr; C, p. 595; G II, p. 277.

principle applies to all things in general;[11] here the influence of the Cartesian usage of the term is, I think, strong.

The new physics and its revolutionary view of motion led Descartes and Hobbes to redefine the notion under scrutiny. Now, according to the Aristotelians there is, of course, a natural terminus, or *telos*, to all change, motion included, and things strive to actualize it. According to this kind of teleology, for instance, heavy inanimate bodies such as rocks strive to move downwards and attain their natural position by reaching the earth. For example, Aquinas writes:

> [I]t is not possible for agents to proceed to infinity: and consequently there must be something, which being attained, the efforts of the agent cease. Therefore every agent acts for an end. (*SCG* III.2)

The end is 'that wherein the appetite of the agent or mover is at rest' (*SCG* III.3). So, and as Aristotle's *On the Heavens* (277a13–27; *CWA* I, p. 460) already makes explicit, since all natural change has both a starting point and a goal, there cannot be (rectilinear) motion that continues to infinity.[12]

The novel conception of motion introduced by Galileo caused, gradually, the demise of the teleological world-view and an upheaval in the natural sciences. The core of the Galilean conception was formed by the explanation model for the persistence of motion that was later to be known as the principle of inertia. According to it, any moving body continues its rectilinear motion unchanged *ad infinitum* if not interfered with (by something external to the body). Descartes's famous first law of nature states:

> [E]ach and every thing, insofar as it is in itself [*quantum in se est*], always continues in the same state; and thus what is once in motion always continues to move. (*PP* 2.37; CSM I, p. 240, translation modified)[13]

The difference between this formulation and for instance that of Aquinas cited above (*SCG* III.2) is striking. In the early modern framework, changes do not have natural termini, and inertial motion goes on for ever.[14] For our purposes the main thing to note is that when such thinkers as Descartes and

[11] Cf. Wolfson 1961 (1934) II, p. 199; Oittinen 1994, p. 180. It seems in fact that in this respect Spinoza's *conatus* resembles more another Stoic notion, namely that of *pneuma*: as A. A. Long (2003, p. 374) observes, noting the analogy to Spinoza's *conatus*, '[t]he Stoics explained the persisting identity of particular beings by reference to the "sustaining" power of the thing's internal *pneuma*' with which inanimate things, too, are endowed. For more on *pneuma* as an active principle that permeates material objects, see Løkke 2009.

[12] To be exact, this point does not apply to circular motions natural to the celestial realm; see Gaukroger 2006, p. 187.

[13] Further, as *PP* 2.39 (CSM I, p. 241) makes clear, 'all motion is in itself rectilinear'.

[14] Carriero 2005, p. 121; see also Hoffman 2009, p. 137.

Hobbes offered philosophical elaborations of this idea, they chose to explicate the nature of motion in terms of striving or endeavour.

Descartes explains in his *Principles of Philosophy* that

> [w]hen I say that the globules of the second element 'strive' to move away from the centres around which they revolve, it should not be thought that I am implying that they have some thought from which this striving [*conatus*] proceeds. I mean merely that they are positioned and pushed into motion in such a way that *they will in fact travel in that direction, unless they are prevented by some other cause* (*PP* 3.56; CSM I, p. 259, emphasis added),

and 'depending on the causes we are considering, we may say that the body is tending or striving to move in different directions at the same time' (*PP* 3.57; CSM I, p. 259). In other words, '*conatus*' is a disposition term, designating a tendency or inclination: if not interfered with, things will continue to move as they presently do. Moreover – and here we may recall the discussions concerning the Cartesian problem of power – Descartes defines finite things' power in terms of tendency:

> [W]e must be careful to note what it is that constitutes the power [*vis*] of any given body to act on, or resist the action of, another body. This power consists simply in the fact that everything tends, insofar as it is in itself [*quantum in se est*], to persist in the same state, as laid down in our first law. (*PP* 2.43; CSM I, p. 243, translation modified)

In other words, Descartes identifies the power and *conatus* of a body with the tendency to move in a given direction or in general to persist in the same state, insofar as the body is in itself.

However, things are more complicated than they may first appear, for as the above-cited passage also indicates, Descartes wants and needs something stronger than this, namely something on which can be grounded the fact that when things encounter opposition, they do not simply cease to be in their present state but *resist* the opposing factors. This is made more explicit in the earlier *The World* in which Descartes says that when a body tends to move in a certain direction, this means only that 'it is disposed to move there, whether it actually moves or whether some other body prevents it from doing so'; to this Descartes adds, importantly, that 'it is principally in this last sense that I use the word "tend" [*tendre*], because it seems to signify some exertion [*effort*] and because every exertion presupposes some resistance' (*World* XIII). In *PP* 2.43 (CSM I, p. 241), he moves similarly from *tending to persevere in one's state* to *resisting change*. But these moves do not provide much of an argument; to make things worse, it is hard to see how one could be extracted out of Descartes's conceptual framework. In a letter to de Volder, Leibniz insightfully expresses the difference between the mere tendency to continue in the prevailing state and the power to resist change:

> But even if there is a force in matter for preserving its state, that force certainly cannot in any way be derived from extension alone. I admit that each and every thing remains in its state until there is a reason for change; this is a principle of metaphysical necessity. But it is one thing to retain a state until something changes it, which even something intrinsically indifferent to both states does, and quite another thing, much more significant, for a thing not to be indifferent, but to have a force and, as it were, an inclination to retain its state, and so resist changing. (AG, p. 172)

In light of this well-founded distinction, Descartes's reasoning seems to involve an illegitimate move from identifying power with a mere indifferent tendency to claiming that tendency implies resistance.[15] Thus, even if a convincing account were given of the way in which powers as tendencies pertain to extended bodies, there would still remain the problem of showing how to get from tendencies to resistance. This is what I would call the Cartesian problem of resistance.[16] Once again, Descartes is caught in a dilemma: he decisively weakens (arguably to keep his system clear of everything that might even hint at occultness) the notion of power into a mere tendency to continue, if not interfered with, in the prevailing state; but at the same time, he thinks that this kind of *conatus* licenses the idea of resistance. It seems to me that Descartes cannot have it both ways.

Hobbes's account of striving is also an important early modern precursor of the Spinozistic doctrine, and his way of basing his influential political theory on the naturalistic principle of endeavour no doubt attracted Spinoza's attention.[17] As perhaps the clearest example of a mechanistically oriented thinker, the concept of motion forms the very core of Hobbes's thought; as Doug Jesseph maintains, 'Hobbes conceived of motion as the most basic explanatory concept in all of philosophy, so his account of first philosophy must be framed in terms of motion and its principles.'[18] Consequently, we should not be surprised to discover that Hobbes defines striving – or as he calls it, endeavour – in terms of motion, not power:

[15] On the basis of Leibniz's passage, Garber (1994, p. 48) correctly maintains: '[F]rom the simple tendency to remain in a given state, one cannot infer that there exists a force, a tendency, or conatus *opposing* that change and keeping a body in the same state. And he [Leibniz] is right. But, and it is this that I want to emphasize, this is exactly what Descartes infers.' See also Des Chene 1996, pp. 311, 339–41.

[16] For more on this issue, see Pietarinen 2001, p. 75.

[17] On this, see Curley 1988, pp. 104–8, 119–26. For a very instructive account of Hobbes's conception of power and endeavour, see Pietarinen 2009b.

[18] Jesseph 2006, p. 129.

I define ENDEAVOUR *to be motion made in less space and time than can be given*; that is, *less than can be determined or assigned by exposition or number*; that is, *motion made through the length of a point, and in an instant or point of time*. (*De Corpore* III.15.2; *EW* I, p. 206)

In *Leviathan* he writes:

[T]hese small beginnings of motion, within the body of man, before they appear in walking, speaking, striking, and other visible actions, are commonly called ENDEAVOUR. (*Leviathan* I.6; *EW* III, p. 39)

So, 'endeavour' refers to infinitely small beginnings of motion that result, eventually, in activities of different kinds, most importantly self-preservation: '[T]he first foundation of natural right is this, that *every man as much as in him lies endeavour to protect his life and members*' (*De Cive* I.7; *EW* II, p. 9). However, since 'endeavour is to be conceived as motion' (*De Corpore* III.15.2; *EW* I, p. 206),[19] it is unclear how it could ever be used to *explain* either inertial motion or resistance, both of which can be found in Hobbes's theory: just saying that unhindered motion continues for ever because of very small beginnings of motion is not very helpful, and more than contrariety of motions seems to be required for a proper concept of resistance.[20] Accordingly, despite Hobbes's exemplary status for Spinoza, his notions of power and, consequently, endeavour seem too weak to have any real explanatory force.[21] This, I think, is the price Hobbes pays for not being prepared to include in his ontology anything but corporeal bodies in motion.

So there was a well-established tradition of *conatus*-related thinking in Western philosophy: the Stoics and the Aristotelians endorsed some version of the view that endeavouring and desiring are basic to at least all animals and that this kind of conation is especially connected to promoting self-preservation. Early modern thinkers, for their part, stripped the idea of striving of its anthropomorphic overtones while trying to solve the philosophical questions raised by the new natural science and its conception of motion.[22] Such a prominent figure as Hobbes even founds his theory of psychology and politics on the principle of striving for self-preservation, thus paving the way for the Spinozistic expansion of the idea of striving

[19] Cf.: '[A]ll active power consists in motion' (*De Corpore* II.10.6; *EW* I, p. 131).

[20] 'I define RESISTANCE *to be the endeavour of one moved body either wholly or in part contrary to the endeavour of another moved body, which toucheth the same*' (*De Corpore* III.15.2; *EW* I, p. 211).

[21] I thus completely agree with Jesseph 2006, pp. 150, 152; see also Pietarinen 2001, p. 80; 2009b.

[22] Cf.: 'Motion has a force [*vim*] of persevering in its state; this force is really nothing other than the motion itself – that is, the nature of motion is such' (*CM* I.6; C, p. 314; G I, p. 248).

from the sphere of motion of bodies to the being of all things. However, the simple fact that 'conativist' theses abounded in and before Spinoza's time is surely not, in itself, a sufficient reason for a critical thinker, ready to allow only carefully scrutinized elements in his system, to adopt the notion. Spinoza's version of the doctrine was no doubt influenced by many sources, but we should still ask, why is the doctrine located in the *Ethics*? Or: the importance of the notion of power has already been acknowledged, and it is, evidently, connected to the notion of *conatus* – but why should power be, at least in its one fundamental form, *striving against opposition*?

At this point, a certain general feature of the seventeenth-century philosophy of physics proves to be crucial. Alan Gabbey articulates it very well:

> Taking seventeenth-century 'dynamics' as a whole, insofar as this is permissible, it can be said that the great majority of its practitioners understood force in its functional sense as that concomitant of a body – expressed in terms of its whole speed and corporeal quantity – which could be identified with the body's relative capacity to overcome a similarly understood resisting force, whether potential or actual, irrespective of the speed and corporeal quantity in terms of which the contrary force was expressed. *Interactions between bodies were seen as contests between opposing forces*, the larger forces being the winners, the smaller forces being the losers.[23]

Gabbey calls this *the 'contest' view of force*.[24] Descartes may be seen as espousing it; the following passage from the *Principles of Philosophy* is noteworthy:

> To enable us to determine, in the light of this, how individual bodies increase or diminish their motions or change direction as a result of collision with other bodies, all that is necessary is to calculate the power of any given body to produce or resist motion; we also need to lay it down as a firm principle that the stronger power always produces its effects. (*PP* 2.45; CSM I, p. 244)

This kind of view of the interactions of bodies as contests between forces forms an important part of Spinoza's intellectual milieu. As Carriero notes, in a plenary universe like Descartes's and Spinoza's, 'things are continually getting in one another's way',[25] so also from this point of view it is understandable to see the natural world as a field of contest. We find mechanistically and reductionistically bent Hobbes echoing this rather violently as he writes in his *Elements of Law* that human life is comparable to a race in which one can for instance 'break through' and 'lose ground' – a race having 'no other *goal* [. . .] but being foremost' (*EW* IV, pp. 52–3). I have argued that in Spinoza's essentialism things have power to cause effects

[23] Gabbey 1980, p. 243, emphasis added. To this Gabbey adds that this is 'a conception of evidently anthropomorphic origin'. However, I am uncertain of why it would be that.
[24] Ibid., p. 244. [25] Carriero 2005, p. 133.

through their essences; this power is always fully exercised, only the extent to which its exertion results in actions and passions varies. Now, the just-sketched vision of finite things' existential conditions, together with the groundbreaking idea that motion has no natural termini, has an impact on how power must be conceived, and precisely these are explicated by the *conatus* doctrine that elucidates the nature of essential causers in the temporal field of resistance and aid. The very proof of the *conatus* proposition, to which we will now turn, gives important indications of the way in which this happens.

SOME INFLUENTIAL INTERPRETATIONS OF THE *CONATUS* ARGUMENT

The proof of 3p6 runs as follows:

For singular things are modes by which God's attributes are expressed in a certain and determinate way (by 1p25c), i.e. (by 1p34), things that express, in a certain and determinate way, God's power, by which God is and acts. And no thing has anything in itself by which it can be destroyed, *or* which takes its existence away (by p4). On the contrary, it is opposed to everything which can take its existence away (by p5). Therefore, as far as it can, and it is in itself [*quantum potest, et in se est*], it strives to persevere in its being, q.e.d. (3p6d, translation modified)

So, Spinoza cites 1p25c, 1p34, 3p4, and 3p5 in his argument. Given that they are all relatively concise, they can be quoted in full, with their demonstrations, in order of appearance:

Particular things are nothing but affections of God's attributes, *or* modes by which God's attributes are expressed in a certain and determinate way. The demonstration is evident from p15 and d5. (1p25c)

God's power is his essence itself. (1p34)

For from the necessity alone of God's essence it follows that God is the cause of himself (by p11) and (by p16 and p16c) of all things. Therefore, God's power, by which he and all things are and act, is his essence itself, q.e.d. (1p34d)

No thing can be destroyed except through an external cause. (3p4)

This Proposition is evident through itself. For the definition of any thing affirms, and does not deny, the thing's essence, *or* it posits the thing's essence, and does not take it away. So while we attend only to the thing itself, and not to external causes, we shall not be able to find anything in it which can destroy it, q.e.d. (3p4d)

Things are of a contrary nature, i.e., cannot be in the same subject, insofar as one can destroy the other. (3p5)

For if they could agree with one another, or be in the same subject at once, then there could be something in the same subject which could destroy it, which (by p4) is absurd. Therefore, things etc., q.e.d. (3p5d)

There is a considerable amount of disagreement over this argument. Many commentators have contended that only some, or even just one, of its ingredients do real work in the argument. Recording everything that has been said about this derivation would, were it possible, not make sense; so I will only generally delineate those discussions that deal directly with the question of how the derivation works. Naturally, they form the indispensable background against which I will develop my own views.

It is helpful to note that in the literature there have been, roughly, two different types of approach to the *conatus* argument. The dominant one has emphasized the conceptual discussions included in the latter part of the derivation, i.e. 3p4 and 3p5, according to which no thing can destroy itself or contain anything self-destructive; the striving to persevere in one's being is supposed to follow from this. The *locus classicus* of modern *conatus*-criticism, Bennett's *Study*, proceeds along these lines,[26] and much of the subsequent discussion has followed its lead. Most importantly, Bennett succeeds in locating what is, at least in this approach, by far the weightiest problem in Spinoza's argument: 3p6 can be interpreted as saying, *any thing exerts itself against anything destructive*, but it appears simply impossible to derive something this strong from the mere *any thing is unlike anything destructive* of 3p5.[27] In other words, if 3p6 is derived from 3p4 and 3p5 alone, obviously no answer can be given to the question, where does a focal element of the *conatus* principle, resistance to opposition, come from?[28] Viewed in this light, the argument seems to be, as Bennett claims, fallacious.[29]

Another type of approach to the *conatus* argument, less popular but nevertheless significant, has been to emphasize the material located in the

[26] Note that Bennett has not budged from this position; see Bennett 2001, p. 218.

[27] Bennett 1984, p. 242.

[28] Cf. Garber 1994, pp. 60–2, 64. For attempts to defend Spinoza, see Curley 1988, p. 109; Manning 2002, pp. 185–7; Pietarinen 2003, p. 143.

[29] Della Rocca's position, which focuses on 3p4, is especially interesting. In his 'Spinoza's Metaphysical Psychology' (1996, pp. 200–6) he claims 3p6 to be derived from 3p4 alone and thinks that the derivation involves a conflation of two different readings of 3p4; moreover, it is difficult to see how Della Rocca's approach here could yield genuine resistance against opposition. However, in his recent *Spinoza*, Della Rocca (2008b, pp. 137–47) has a much more positive view on the derivation: he sees the principle of sufficient reason as the driving force behind it and thinks that 3p6 does follow from 3p4. Moreover, he addresses (pp. 150–3) the issue of resistance, arguing that Spinoza reduces causation to conceivability or conceptual dependence (which reduction also results, according to Della Rocca, from the use of the principle of sufficient reason, see esp. p. 45) and thus, given that there is for Spinoza a conceptual connection between the cause and the effect – to present this in Della Rocca's terms, between e.g. the motion of a rock and the breaking of a window so that we can see that the rock will break the window unless other things intervene – Della Rocca argues that Spinoza can say that the cause exercises power, or strives against intervening factors, even in cases in which this striving is unsuccessful. For my criticism of this view, see n. 46 below.

beginning of 3p6d, i.e. 1p25c and 1p34. The key idea here is that if things are proven (by 1p25c and 1p34) to be active or intrinsically powerful, the *conatus* principle follows; the conceptual examinations directly preceding the demonstration are regarded as subsidiary or somehow preliminary, if even that.[30] Now, it is a good idea to take the beginning of the demonstration seriously because it grounds, as I will explain later on, the notion of resistance to opposition; but an informative reading should give us a balanced view of the relationship that 1p25c and 1p34 have with 3p4 and 3p5, and as far as I can see, no account of this type quite succeeds in this.[31]

It seems, then, that a fully satisfactory interpretation of Spinoza's argument still remains to be given. As it is my contention that 3p6d is basically valid and contains no idle elements, the challenge is to offer an enlightening reading of the derivation that shows how, exactly, it is supposed to work. My analysis is, of course, more in line with some of the previous accounts than with others, and so I will situate it in the context of the two types of approach and explicate some of the more notable differences between my views and the ones presented in this section.

THE IMPOSSIBILITY OF SELF-DESTRUCTION

Since Spinoza has so often been regarded as trying to derive the *conatus* doctrine from 3p4 and 3p5 alone, we can begin by analysing the latter part of 3p6d:

30 Matheron (1988 [1969], pp. 10–11) and Henry Allison (1987, pp. 131–3) read Spinoza along these lines. Drawing on Matheron, Allison (1987, p. 133) claims 3p6 to be a 'reformulation' of 3p4 and 3p5 in positive terms: 'Since things act, and since [. . .] they cannot act in ways that are contrary to their nature – that is, which tend to their self-destruction', the *conatus* principle results. Moreover, 'insofar as a thing acts, this opposition to whatever tends to destroy it is expressed as an actual resistance; and [. . .] for a thing to act in such a way as to resist whatever tends to destroy it is to act in a self-maintaining way' (p. 134). However, Matheron and Allison write on this topic very briefly, and it is difficult to say what is the precise meaning of the suggested reformulation. More recently, Lin (2004, pp. 25–43) has argued for the second approach. According to him, Spinoza offers not just one but two separate arguments in favour of the *conatus* principle: an unacceptable one based on 3p4 and 3p5, and a valid one based on 1p25c and 1p34. The good argument is grounded on the expressive relationship that obtains between finite things and God's power: 'Because our actions express the divine power whereby God creates everything he can, we too must strive to do everything in our power' (Lin 2004, p. 42). Although I do not think that 3p6d contains two arguments, Lin surely is right in emphasizing the notions of power and expression, both topics for further discussion below. See also Curley 1988, pp. 112–13; Schrijvers 1999, p. 71; Steinberg 2005, p. 154; Schmidt 2009, p. 301; LeBuffe 2010, p. 103.

31 Garrett's (2002, pp. 136–46) outlook on the matters at hand definitely also merits attention. It is somewhat difficult to say how his position relates to the two types of approach presented above, and this stems, I believe, largely from the fact that according to Garrett, the material cited in 3p6d is there to show *how* the doctrine is true – *that* it is true already follows from Spinoza's views on inherence, conception, and causation, which are stated in the beginning of the *Ethics* (1d3, 1d5, 1a1, and 1a4). Although Garrett's discussion contains a wealth of important ideas, I find it problematic assigning this kind of auxiliary or confirmatory role to 1p25c, 1p34, 3p4, and 3p5; see n. 50 below.

And no thing has anything in itself by which it can be destroyed, *or* which takes its existence away (by p4). On the contrary, it is opposed to everything which can take its existence away (by p5). Therefore, as far as it can, and it is in itself [*quantum potest, et in se est*], it strives to persevere in its being, q.e.d. (3p6d, translation modified)

Spinoza considers 3p4 self-evident; but at least for us, the proposition is far from obvious, so there is some interpretative distance to be travelled.

A quick look at 3p4 and 3p5 reveals that they are based on Spinoza's theory of definitions and essences. This is not the only place where it makes its presence felt: as I propounded when examining issues pertaining to being and causation, Spinoza accepts a view, not uncommon in his day, that each thing has both an essence and a definition that captures that essence; according to his geometrized version of this view, each thing has both (1) a definition that expresses the thing's essence; this definition not only states how the thing in question is produced but also what properties it would have on the basis of its essence alone, and (2) an essence from which, once the thing is instantiated, those properties necessarily follow or flow insofar as the thing in question is unaffected by external causes. Having recalled this, we can return to the demonstration of 3p4:

[T]he definition of any thing affirms, and does not deny, the thing's essence, *or* it posits the thing's essence, and does not take it away. So while we attend only to the thing itself, and not to external causes, we shall not be able to find anything in it which can destroy it. (3p4d)

Although there has been much discussion on what Spinoza here means by 'the thing itself',[32] the reference to definitions and essences makes it clear, I think, that he is referring to a thing as it would be constituted by its definable essence alone, completely uninfluenced by other things or 'external causes'; moreover, this would be in line with the appearance of '*quantum in se est*' in 3p6 itself, a phrase that was traditionally taken to refer to what takes place by a thing's nature or essence.[33] I have already observed that essences individuate things, make them what they are, indicate the manner in which substance must be modified for a definite thing to exist, or – to use Spinoza's brief expression – 'posit' things (2d2). Given this, it would be hard to claim that an essence could include anything capable of taking away the thing whose essence it is; this would only go to show that what we had was no true essence to begin with. Correspondingly, a definition states how the

[32] This discussion dates from Bennett's *Study* (1984, pp. 236–7); cf. especially Garrett 2002, pp. 130, 147.

[33] See Ch. 3 above. As Lee Rice has pointed out to me, also 4p20s gives support for this reading; see also Barbone and Rice 1994, p. 235.

definiendum can be produced; so including something destructive to the *definiendum* goes against the very idea of a proper definition. Accordingly, in the Spinozistic scheme of things 3p4 seems quite secure,[34] and the approving remarks made by many commentators reflect this fact. A passage from *Behind the Geometrical Method* can serve as an example:

> We must imagine the definition of a thing as a formula which describes a process by which a thing of that kind might be produced, as stating conditions which would lead to the existence of a thing of that kind, i.e., of a thing having all the properties common and peculiar to that kind of thing. So as long as we focus on that formula, we will, of course, find nothing which would entail the non-existence of the thing.[35]

And it surely seems that Spinoza would find it unthinkable that any such 'formula' would entail, at any point in time, precluding the existence of the entity whose formula it is, because that would only imply that the definition is no true definition at all, making the *definiendum* a non-thing. Spinoza thus seems to be here emphasizing the first requirement he sets for a proper definition (*TdIE* § 96; C, pp. 39–40), that it must state how the defined thing is to be produced.

3p5, in turn, reads:

> Things are of a contrary nature, i.e., cannot be in the same subject, insofar as one can destroy the other. (3p5)
>
> For if they could agree with one another, or be in the same subject at once, then there could be something in the same subject which could destroy it, which (by p4) is absurd. Therefore, things etc., q.e.d. (3p5d)

So the claim is that mutually destructive items are of a contrary nature and cannot be in the same 'subject' (*subjectum*), because this would violate the just presented 3p4. The concept of subject is puzzling; it appears only twice in the *Ethics* and is never defined. However, the fact that 3p4 is allowed to restrict what may inhere in a subject suggests that here, too, we find Spinoza's theory of definitions and essences at work. Thus, I take it that by subject Spinoza means a thing as it would be constituted solely by its definable essence, with only those properties that can be derived from its proper definition. Here I am in agreement with Garrett who contends:

> For Spinoza, an individual or singular thing exists to the extent that there is instantiated a definite essence or nature that can serve as a locus of causal activity.

[34] It would also have been traditionally regarded as such; cf. Des Chene 1996, p. 71.

[35] Curley 1988, p. 111. See also Matheron 1988 (1969), p. 10; Allison 1987, pp. 131–3; Donagan 1988, pp. 149–51; Garber 1994, pp. 59–60; Youpa 2003, p. 481; Lin 2004, pp. 26–7; Nadler 2006, pp. 196–7; Della Rocca 2008b, pp. 137–43. For a differing view, see Matson 1977, pp. 406–8.

Where there is such an essence, properties follow (both causally and logically) from that essence, and hence one can speak of a 'subject' in *which affections exist.*[36]

That 'subject' is here given a special essentialist sense is brought forth by Garrett's *prima facie* surprising-sounding claim that to the extent qualities of things are produced by external causes and thereby not conceived solely through the subject of which they are predicated, they do not inhere in that subject.[37] Thus, in 3p5 Spinoza seems to be highlighting the other requirement for a proper definition (*TdIE* § 96; C, p. 40), that from it must be derivable all those properties that follow from the *definiendum*'s essence alone.

Given the aforesaid, how should 3p5 be read? Now, if the necessary properties derivable from an alleged definition involved logical opposition, the *definiendum*'s essence would involve contradiction and thus be self-negating. But this is precisely what 3p4 precludes. So, for instance, from no proper definition can be deduced both the property of having a hypotenuse whose square equals the sum of the squares on the other two sides, and the property of having the sum of the internal angles equalling that of four right angles, because the first property implies the negation of the second. In other words, granted the theory of definitions and essences underlying Spinoza's view on subjecthood, it is well-founded to claim that no subject can involve anything self-destructive, i.e. contradictory, for that would only prove that we did not have a true thing or subject, with a definable essence, to begin with.[38] Recall Chapter 1 and the claim that each thing has its specific internal structure; on this view, a subject with contradictory properties would, in fact, amount to the repugnant situation in which there would be something irrational and incomprehensible in the way reality itself is built. Consequently, Spinoza feels himself entitled to assert, '[t]hings are of a contrary nature, i.e., cannot be in the same subject, insofar as one can destroy the other'.

Unfortunately all this does not mean that Spinoza would be out of trouble. For as Kant (*NM* § 1) with his customary brilliance and instructiveness teaches us, a distinction between two kinds of opposition should be kept in mind: (a) *logical* opposition through contradiction (when something is simultaneously affirmed and denied of the same thing), having the consequence of an unthinkable and impossible nothing, and (b) *real* opposition in which two properties are opposed and cancel out each other, but without contradiction; and in this latter kind of opposition, the conflicting determinations can exist in the same subject (for instance, a

[36] Garrett 2002, p. 150, emphasis added. [37] Ibid., p. 140.

[38] Thus I would agree with Steven Nadler (2006, p. 196).

body may have a certain motive force to go left and a force of same size to go right, resulting in the state of rest, but not in the impossibility of the body).

Now, Spinoza's way of using 3p5 in 3p6d (any thing 'is opposed to everything which can take its existence away [by p5]') obviously refers not to logical but to real opposition. This is only to be expected given the fact that *conatus* is a form of power, and oppositions of powers are real, not logical in character. So, if 3p5 is taken to mean that from no definition can items in logical contradiction be derived, there is no way to extract the opposition thesis of 3p6d out of 3p5 alone, for a completely different sort of opposition pertains to 3p6 than to the proposition preceding it. As I believe this to be the source of Bennett's complaint that Spinoza illicitly moves from mere difference to exertion against opposition, this is no minor problem. I will later present some suggestions on how it might be solved, but due to this defect, 3p6d seems to be unsatisfactorily formulated.

However, I do not think that Spinoza is trying to derive the *conatus* doctrine from 3p4 and 3p5 alone, so the fact that that cannot be done is not, as such, perilous. Instead, we need a proper interpretation of the role played by the material cited at the beginning of the proof, which is what I will try to offer next. I believe that a firm grasp of 1p25c and 1p34 enables us to see 3p4 and 3p5 in a new light and to discern how the whole derivation is supposed to work.

POWER AND EXPRESSION

The *conatus* argument begins as follows:

> For singular things are modes by which God's attributes are expressed in a certain and determinate way (by 1p25c), i.e. (by 1p34), things that express, in a certain and determinate way, God's power, by which God is and acts. (3p6d)[39]

1p25c, in turn, states:

> Particular things are nothing but affections of God's attributes, *or* modes by which God's attributes are expressed in a certain and determinate way. The demonstration is evident from p15 and d5.

And the by now familiar, very concise 1p34:

> God's power is his essence itself.

[39] '*Res enim singulares modi sunt, quibus Dei attributa certo, et determinato modo exprimuntur (*per Coroll. Prop. *25.* p. *1.), hoc est (*per Prop. *34.* p. *1.*) *res, quae Dei potentiam, qua Deus est, et agit, certo et determinato modo exprimunt*' (G II, p. 146).

The general structure of this part of the argument is clear enough: modes express attributes in a certain and determinate way (1p25c), attributes constitute or express God's essence (an implicit premise, 1d4, 1d6, 1p10s), and God's essence is his power (1p34); thus, particular things as finite modes express in a certain and determinate way God's power.

We can consider 1p34 first. Our discussion thus far, especially in Chapter 3, reveals that citing this proposition emblematic of Spinoza's dynamic essentialism can only mean that the roots of the *conatus* doctrine run deep, right to those opening propositions of the *Ethics* that deal with the basics of existence and causation (I would thus object to the claim that 3p6 is not really connected to the material preceding it). With regard to the *conatus* argument, all this suggests that the appearance of 1p34 in 3p6d can be interpreted as an extremely economical reference to a general metaphysical position, unearthable from the first part of the *Ethics*, in which physical and mental phenomena are caused by the action of particulars endowed with power to cause effects according to their essences.[40]

Understanding 1p25c, 'things are [. . .] modes by which God's attributes are expressed in a certain and determinate way', requires a fair amount of unpacking as well. It includes once again a notion Spinoza nowhere defines, that of expression. For our purposes the crucial point is, to my mind at least, that in all the contexts in which it occurs a central feature of the expressive relationship is that if y expresses x, y is, of course, in some way different from x, but still in such a manner that y *retains* or *preserves* the basic character or nature of x. Substance has many attributes, none of them simply equivalent with the essence of substance, but as all attributes are faithful to that essence in their diverse ways of constituting it, they can be said to express it. Further, finite things express their attribute, because they are manners in which a certain attribute is modified, and as such, of course, their basic nature is that of their attribute.[41]

The idea of expression is closely connected to that of immanence: Spinoza claims the fact that things are expressions of God's attributes (1p25c) to be evident from 1p15, the overtly immanentist proposition proclaiming everything to be in God. Gilles Deleuze stresses this connection; his point in its general outline seems to be that unlike in traditional theology with a transcendent God whose being differs, due to the

[40] Cf. Cohen (1964, p. 148): an 'essential part of the sense of *quantum in se est*' was in the seventeenth century taken to be 'the idea of "naturally" or "by its own force"'. On this, see Ch. 3 above.

[41] For informative recent discussions on expression in Spinoza's system, see Lin 2004, esp. pp. 36, 45, 2006, pp. 342–3.

impassable ontological gulf located between God and us, radically from ours, in Spinoza's immanent system all being is univocal:

> Thus all imitative or exemplary likeness is excluded from the relation of expression. God expresses himself in the forms that constitute his essence[.] [. . .] [O]nly univocal being, only univocal consciousness, are expressive. Substance and modes, cause and effects, only have being and are only known through common forms that actually constitute the essence of the one, and actually contain the essence of the others.[42]

Thus, Spinoza's concept of expression involves the denial of finite things as creations of transcendent God, radically separate from and only somehow analogically resembling their creator. There is only one existent, substance whose essence is constituted by attributes, and finite things are expressions of those attributes. For instance, each finite body is a way in which the attribute of extension is modified, and thus they cannot but express that attribute (recall also the geometrical illustration of 2p8s).

The phrase 'in a certain and determinate way' occurs quite often in the *Ethics*, but its exact meaning is not completely evident. In his letter to Hudde, Spinoza says that the notion of determinate 'denotes nothing positive, but only the privation of existence of that same nature which is conceived as determinate' (*Ep*36; S, p. 858; G IV, p. 184). Thus, 'determinate' refers to limitedness; whereas 'certain' would seem to denote particularity or specificity.[43] As a consequence, I would suggest that 'certain and determinate' means 'particular and limited'. So, 1p25c says that finite things are of the basic character of their attribute, albeit in a particular and limited mode.

Taken together, 1p25c and 1p34 imply, in Spinoza's idiom, that things 'express, in a certain and determinate way, God's power, by which God is and acts'. We should now be able to better understand the import of this claim: any singular thing expresses God's power, because singular things, while retaining what is characteristic of God's power, modify that power in a particular limited fashion. This, I think, exemplifies Spinoza's dynamic essentialism according to which finite things are specifically determined loci of (God's) power, not distinguishable from a specific manner of operation and causation. Since individual essences determine how substance and its causal efficacy is modified, it can be said that finite things express God's power because it is, as it were, distributed through finite essences in

[42] Deleuze 1992 (1968), p. 181.

[43] Hence I would largely agree with Charles Ramond (1995, p. 78), according to whom 'certain' means, in this context, 'precise'.

actuality, and the frame of distribution changes from one moment of time to another while God-or-Nature's total power stays infinite.

THE ARGUMENT RECONSTRUCTED

How should the *conatus* argument be evaluated in light of the present interpretation? Evidently, its beginning, built on 1p25c and 1p34, brings forth a certain dynamic framework in which finite things are centres of causal power, capable of producing effects in virtue of their essences. Now, the way I see it, this makes the beginning of the demonstration irreplaceable – after all, *conatus* is one form of power, and 1p25c and 1p34 not only bring the notion of power into play, but also inform us on how finite things' power should be understood in the monistic system. So, I disagree with those commentators who see Spinoza as trying to derive the *conatus* doctrine from 3p4 and 3p5 alone; and I agree with those who stress the importance of 1p25c and 1p34. But this raises new questions: why are 3p4 and 3p5 needed at all? Could 3p6 not be derived from 1p25c and 1p34 alone, as Martin Lin argues?[44]

A close look at what is at stake in 3p6 suggests that the answer to this question must be negative. For 1p25c and 1p34 imply only that finite things are, in essence, dynamic causers; but this is not enough to guarantee that they could not act self-destructively or restrain their own power, which would make them incapable of self-preservation. However, this would go against 3p4, and Spinoza uses it to claim, 'no thing has anything in itself by which it can be destroyed, *or* which takes its existence away'. So thus far he has proven that finite things are coherent causers, that is, entities endowed with power and, insofar as they cause effects solely in virtue of their essence, never use their power self-destructively.

Evidently, Spinoza does not think this to be enough. The significance and role of the final plank in the demonstration, 3p5, needs still to be determined. Indeed, considering its content and the way it is used in the demonstration, it seems to me to be a surprisingly decisive ingredient in the argument. Namely, that what 'each thing, insofar as it is in itself', that is, insofar as any thing is considered disregarding everything external to it, strives to preserve, is its *being* (*esse*). However, in the earlier *Theological-Political Treatise* the *conatus* principle is formulated as follows: '[I]t is the supreme law of Nature that each thing strives to persist in its state [*in suo statu*], as far as in it lies, taking account of no other thing but itself' (*TTP* XVI; S, p. 527; G III, p. 189,

[44] See n. 30 above.

translation modified). The only notable difference is thus that in 3p6 the 'state' is replaced by 'being'. However, this difference seems to have received little attention, and 'being' is commonly read as meaning, roughly, *existing in the present state*. In fact, the proposition's reference to being can be regarded as something of a puzzle given that the notion is not to be found in *any* of the material cited in the demonstration. But if we keep in mind the view of subjecthood 3p5 refers to, together with the general doctrine of being discussed in Chapter 1, we can see what kind of 'being' or 'existence' is meant in 3p6. For Spinoza, each subject has a definable essence from which, as far as the subject in question is in itself, certain properties or effects necessarily follow;[45] consequently a subject's being involves not only instantiating a certain essence, but also *being of a specific character*, i.e. being as structured by a definable essence – or being that results when an individual essence modifies a divine attribute.

Given the aforesaid I would like to propose the following reading of the claim included in 3p6d that, by 3p5, each thing 'is opposed to everything which can take its existence away': 3p5 is meant to bring forward that things are not merely non-self-destroyers but subjects from whose definitions properties follow; and as Spinoza thinks to have shown (by 1p25c and 1p34) that finite modifications are entities of power, any subject has true power to produce the properties or effects derivable from its definition, which, Spinoza claims, implies opposing everything harmful. In other words, things exercise power as their definition states, according to their definitions, and thus bringing in the idea of things as expressers of power enables Spinoza to convert logical oppositions into real ones.[46]

On the present interpretation, the argument for 3p6 is structured as follows. First, the beginning of the demonstration brings forward the dynamic framework developed in the first part of the *Ethics*. In it, God's power equals God's ability to cause effects in virtue of his essence alone (1p34). Finite things express this power (1d6, 1p25c), and since in expressing the basic character of the expressed is retained, any finite thing's power is of

[45] To my knowledge the only scholars to have similarly – albeit without delving too deep in this particular topic – emphasized the shift from 'state' to 'being' are Matheron (1990, pp. 268–9) and Chantal Jaquet (2004, pp. 63–4). See also Parkinson 1981, p. 5; Rizk 2006, p. 50.

[46] Thus, I do not think that Della Rocca's proposition (see n. 29 above) solves the problem of resistance. Certainly, each genuine thing has a definition, which is a thoroughly intelligible account of what pertains to the thing as far as it is considered alone, but this is not enough to yield a real opposition, a striving against, or a persistent drive to act in a certain way – this only means that the thing is logically or conceptually incompatible with something else. It may thus be said that there is a certain resiliency to temporal existence that does not reduce to the conceivability of the fact that *x will do A unless prevented*; a resiliency which, I would argue, Spinoza grounds in his system's robust form of power.

precisely the same kind as God's power, only in a specific limited mode. In other words, the idea of expression is supposed to guarantee that just as God's power causes effects that follow from God's nature, finite things cause, with their power, effects that follow from their essences; the same model applies to finite and infinite things alike.

God's power cannot encounter any opposition already for the simple fact that nothing but God exists, but as I noted above, finite things do not find themselves in such happy conditions: temporal reality is a field of constant contest, and consequently things in it do not get to exist and to operate in a hindrance-free, 'frictionless' world. Now, it is not, as such, particularly strange to suggest that there is a linkage between power and resistance; and especially given the just presented way of thinking about interactions between bodies, the concept of power was obviously seen to imply that, in case of opposition, things truly *resist* opposing factors with their power. This applies also to the concept Spinoza endorses: were things simply to cease their causal activities when facing obstacles, the concept of power involved would turn out to be extremely feeble, hardly to be counted as a proper concept of power at all. Further, Spinoza frequently equates power and striving without feeling any need to provide a separate argument for this equation,[47] and this together with his intellectual milieu strongly suggests that he sees the notion of striving against or resisting opposition to be inbuilt in a proper concept of power. Recall Leibniz's observation, that it is one thing 'to retain a state until something changes it, which even something intrinsically indifferent to both states does', and totally another thing 'for a thing not to be indifferent, but to have a force and, as it were, an inclination to retain its state, and so resist changing' (AG, p. 172). In other words, the negative fact that no changes occur in unaffected things is something else than, and not enough to explain, the positive phenomenon of resisting harmful affecting factors. I would argue Spinoza to have implicitly realized this and seen opposition to destruction to stem from the fact that God's modifications are powerful strivers; hence the reference to 1p25c and 1p34. In other words, I submit Spinoza's notion of power to imply that if any thing, whether finite or infinite, encounters opposition, it strives against that opposition to cause effects determined by its own essence alone – the claim 'if opposed, will resist' has an impossible antecedent only with regard to God.[48] The beginning of 3p6d thus implies that taken out of causal isolation, any being endowed with power will, in virtue of its power, exert itself against any harmful external causes encountered.

[47] See e.g. 3p7d, 3p28d, 3p54d. [48] Cf. Della Rocca 2008b, pp. 152–3.

Second, 3p4 guarantees that any thing endowed with striving power never strives to destroy itself, but, third, is (by 3p5) a subject from whose definition an array of properties not in logical opposition to each other can be inferred. Finally, given that things are expressers of power, the exclusion of logical opposition amongst a subject's essential effects or properties is converted into real impulsion against opposing factors, i.e. into striving to persevere in the kind of being determined by the subject's essence alone. That is, 1p25c, 1p34, and 3p5 *together* imply that each thing is really opposed to, or exerts itself against, everything destructive.

However, even the present reading cannot save Spinoza from being guilty of, if not an error, at least of slight imprecision in formulating the latter part of 3p6d: he indicates that exertion against follows from 3p5 alone, and, as Bennett has pointed out, this cannot be the case. Fortunately, this is no fatal flaw, for it can be fixed by bringing in the material located in the very same demonstration.

The foregoing discussion shows that of the two main types of approach to Spinoza's argument, the one emphasizing activity and power of things is the preferable one. Matheron's and Allison's position is especially noteworthy:[49] they also claim that the beginning of 3p6d is crucial for the argument and suggest that it can be connected with the conceptual discussions of 3p4 and 3p5. However, their brief accounts are hardly satisfactory; they emphasize the notion of acting instead of power, without adequately explicating how and why activity is connected to resistance, or telling us where the opposition to destruction is supposed to come from; moreover, they do not even attempt to precisely discern the argument's structure, or how its various elements work together. I hope to have shown that my account is more satisfactory at least in these important respects.[50]

So, to sum up, Spinoza argues in the demonstration of 3p6 that each genuine finite thing is, in itself, an expresser of power (1p25c, 1p34) that never acts self-destructively (3p4) but instead strives to drive itself through opponents to determine being – or produce effects – as decreed by the definition of the thing in question (1p25c, 1p34, and 3p5). Therefore, '[e]ach

49 See n. 30 above.

50 Moreover, Matheron (1988 [1969], pp. 11, 22), too, thinks that 3p6d involves a transition from logical (or conceptual) opposition to physical opposition (which of course is a form of real opposition), but does not explain what could license it. For further criticism against Matheron, see Lin 2004, pp. 26, 28–9. Based on the discussion thus far, I would not be ready to give the material in the official demonstration the kind of role Garrett (2002, pp. 144–6; see n. 31 above) does: most importantly, if I am right, Spinoza sees 1p25c and 1p34 in 3p6d as indispensable for demonstrating the truth of the *conatus* principle, because they bring in the idea of finite things as expressers of power, which, in turn, is his ground for the claim that any thing resists opposition.

thing, insofar as it is in itself, strives to persevere in its being'. The demonstration of 3p6 has its roots deep in Spinoza's ontology, and since its concept of power is supposed to provide the metaphysical grounding for real opposition, the importance of 1p25c and 1p34 should not be underestimated just because Spinoza – as often happens – puts his point exceedingly briefly. Moreover, the derivation is basically valid and contains no superfluous elements. As we will soon see, this interpretation has important implications for how the general nature of the *conatus* doctrine should be understood.

CHAPTER 5

The meaning of the conatus *doctrine*

There are two competing overall readings of the nature and meaning of the *conatus* doctrine. Here the issue of teleology draws the line of division: roughly speaking, one approach is for it, the other against. The significant question is, to borrow Steven Nadler's words, 'whether or not Spinoza, contrary to what would seem to be the lesson of the *Ethics* so far, is surreptitiously and (it has been argued) illegitimately introducing teleology into nature'.[1] It seems that the interpretations presented thus far are relatively unproblematically categorizable within either of the two types of approach. In this chapter I consider an extensive array of interpretations, articulate my reasons for finding the two main interpretative outlooks unsatisfactory despite the fact that there is truth in both of them, and offer my own view of the doctrine on which it appears as a focal facet of Spinoza's geometry of power. I will start by presenting the line of thought in which Spinoza is regarded as the sternest early modern critic of all ends and final causes.

THE INERTIAL READING

There is a time-honoured, and I believe dominant, tradition of viewing Spinoza's system as thoroughly non-teleological in character – a tradition obviously drawing its force largely from the mechanistic tendencies Spinoza displays. Much of the discussion in this section will focus on John Carriero's recent 'Spinoza on Final Causality', because it provides an especially elegant, thorough, and consistent account of this line of interpretation, which I would call *inertial*, and which in its general orientation has been, at least implicitly, quite popular in scholarship. Carriero approaches Spinoza as a proponent of the mechanical philosophy: 'He [Spinoza] presents a theory according to which the sort of causality the new scientists find in the

[1] Nadler 2006, p. 198.

corporeal order is found throughout all of nature.'[2] Moreover, a similar attitude to interpreting Spinoza is explicitly endorsed by such contemporary scholars as Jonathan Bennett and Lee Rice, whose views I will also discuss.

Spinoza's position is, Carriero argues, in sharp contrast to the Thomist conception of finality, in which efficient causes, guided by ends, are needed to actualize perfections. In the early modern scheme of things that Spinoza endorses, changes are not actualizations of potentialities; efficient causes are not end-guided but blind; causal transactions and motion do not have natural termini.[3] What we have are 'pattern-like' beings, or complex systems of matter in motion, that are not internally governed by ends but whose activities are resultants of the tendencies of the simpler parts:

> All of the pattern system's activities are (already?) set by the motive tendencies of its parts (the interaction of those parts with the environment). There is no place in a pattern-like being for a 'form' or 'impression' of the ends of growth and reproduction to shape the exercise of its agency. Moreover, this is true however the system came into being, whether artificially via a clockmaker or naturally through God. The human or divine artificer has set things up so that certain things will likely result in the clock or the plant, so that the clock's hands will move at a constant angular velocity or that over time the plant will, *ceteris paribus*, grow. But after things have been set up, these results eventuate independently of the ends, through motive tendencies blindly following their course.[4]

On this view, *conatus* is a principle of 'metaphysical inertia' derived from the physical principle of inertia:[5]

> The main idea behind Spinoza's account of agency is to take the picture of agency that he finds in simple situations in the new science and to apply it systematically. Simple bodies, when not interfered with by other bodies, just continue to do what they are already doing through a metaphysical inertia. Their motive tendencies are not structured about ends.[6]

In other words, the idea of *conatus* to self-preservation only requires that just as an asteroid once thrust into motion continues that motion for ever (if not impeded), each thing continues to do what it is already doing; exactly as little teleology is involved in any self-preservatory activity as in the case of the asteroid.

When this kind of principle of metaphysical inertia is combined with the view of pattern-like beings with no higher-level organizing principle, the

[2] Carriero 2005, p. 121. See also p. 138. [3] Ibid., p. 121. [4] Ibid., p. 124.

[5] As Carriero (2005, p. 132) points out, 3p6 is formulated in a way reminiscent of Descartes's first law of motion (*PP* 2.37; CSM 1, p. 240). Cf. Rice 1985, p. 249.

[6] Carriero 2005, p. 134. See also Rice 1985, pp. 248–9.

upshot is the claim that complex things, for instance individual human bodies, act the way they do simply because they have motive tendencies 'set by the motive tendencies of their components'.[7] Moreover, on grounds of parallelism it can be claimed that 'adding mentality or consciousness to the picture does not alter the fundamental structure of a motive tendency or appetite'.[8] In other words, the operation of both mind and bodies is determined by causal factors Carriero calls 'motive tendencies'. And it obviously is fully in accordance with this to hold, as Rice does, that '[t]he mental *conatus* is a law of psychodynamic inertia'.[9]

I believe we can outline the emerging picture as follows. The notion of *conatus* refers to a certain kind of metaphysical inertia through which finite things act, but by this nothing more is meant than that there is an attribute-neutral tendency in things to remain as they are. Individuals are just composites formed out of their parts; their behaviour is derived from the numerous differing motive tendencies of their constituents. Under extension, bodies have inertia to continue their prevailing motion for ever, if not interfered with; and, it is argued, we can intelligibly speak about causal transactions under the attribute of thought in a similar fashion: ideas possess mental inertia to stay in existence as they are. The science of mechanics provides the fundamental model in terms of which we should think about thought and other non-extended realms, should there be any.

It is incontestable that Spinoza appeals to desires and appetites in explaining human behaviour, which are defined in terms of the basic *conatus*.[10] This, in turn, has served as the basis for a powerful line of interpretation on which human behaviour is thoroughly non-teleological in character. Bennett has introduced, at least to the current discussion, the claim that Spinoza sees his notion of appetite as the philosophically proper replacement for traditional teleological notions; Spinoza would 'have no truck with the language of final causes unless it is construed in terms of his harmless notion of appetite'.[11] To my mind, as textual support such claims as '[b]y the end for the sake of which we do something I understand appetite' (4d7) are as strong as one could wish, provided that it can be shown that there is nothing teleological involved in our appetites.

Carriero's goal is to present an account of appetites, and thus of human agency, purged of all finality.[12] He begins by providing a reading of a famous

7 Carriero 2005, p. 134. 8 Ibid., pp. 134–5. 9 Rice 1985, p. 249.
10 See especially 3p9s. 11 Bennett 1984, pp. 221–2; see also Parkinson 1981, p. 7.
12 Bennett's (1984, pp. 222–4) argument for analysing appetites in terms of non-teleological intrinsic states turns crucially on the thesis that the representational content of ideas is, for Spinoza, causally

and perplexing claim Spinoza feels entitled to make after defining appetite, will, and desire in terms of the fundamental *conatus*:

> From all this, then, it is clear that we neither strive for, nor will, neither want, nor desire anything because we judge it to be good; on the contrary, we judge something to be good because we strive for it, will it, want it, and desire it. (3p9s)

Carriero takes the idea that we do not want anything because we judge it to be good to follow from the non-end-responsiveness of our motive tendencies.[13] His idea is apparently that if a perceived good were something for which we acted, it would be an end for us; since this would be in conflict with the inertial view, it is something Spinoza cannot allow. The ensuing important point concerning Spinoza's value theory is 'that, contrary to what one might think, the conatus, motive tendencies, [or] appetites [. . .] are prior to our evaluation of something as good. *Our appetites don't track our evaluations; our evaluations track our appetites.*'[14]

After this, Carriero is ready to turn to the Spinozistic account of human action. First, he observes that Spinoza acknowledges us to think of ourselves as directed by ends[15] – which, presumably, explains also Spinoza's occasional apparently teleology-involving talk about ends. However, for Carriero's Spinoza this is a confusion from which we can be rescued by grasping how the true causal architecture of the world works in the case of human agency. The key passage of the *Ethics*, one to which we will return many times below, is the following:

> As he [God, *or* Nature] exists for the sake of no end, he also acts for the sake of no end. Rather, as he has no principle or end of existing, so he also has none of acting. What is called a final cause is nothing but a human appetite insofar as it is considered as a principle, *or* primary cause, of some thing.
>
> For example, when we say that habitation was the final cause of this or that house, surely we understand nothing but that a man, because he imagined the conveniences of domestic life, had an appetite to build a house. So habitation, insofar as it is considered as a final cause, is nothing more than this singular appetite. It is really an efficient cause, which is considered as a first cause, because men are commonly ignorant of the causes of their appetites. (4pr; C, p. 544; G II, pp. 206–7)

impotent (Bennett 1983, pp. 145–7, 1984, pp. 219–20, 2001, pp. 210–14); however, discussing this much-debated thesis (see Curley 1990; Garrett 1999; Manning 2002; Lin 2006) would take us too far afield; for my detailed argument against it, see Viljanen 2010.

13 Carriero 2005, p. 138.

14 Ibid., p. 139, emphasis added. Bennett (2001, p. 215) contends 'the notion of what a desire is *for*' to be got 'out of a likely effect of the desire', but because Bennett obviously sees efficient causation as 'blind', this does not introduce any kind of teleology into Spinoza's thought.

15 Carriero 2005, p. 140.

Carriero expresses his understanding of this passage in such striking words that they deserve to be quoted at length:

> [I]magine that I, somewhat the way we might imagine a bee or wasp, find myself headed toward the nest construction or house building. The way in which I experience this motive tendency is that I find myself being visited by images of domestic bliss and so forth – that's what 'imagin[ing] the conveniences of domestic life' is, for Spinoza – as I go about stacking bricks and so on. Since I do not know where these images come from, I think of them as the first items in a causal chain and come to see the subsequent items in the chain as done for their sake.[16]

However, according to Carriero,

> the images themselves are but links in an enormously complex causal chain that runs parallel to a similarly complex chain of corporeal causes involving my body. If I think of myself as going about the house-building in a purely mechanical way, in the way we might think of a wasp or bee building its nest, and think of the mental side of this activity as running parallel to the corporeal side, I will come to realize that the house-building appetite is in fact an enormously complicated tendency that involves a vast number of more subtle tendencies or urges. As these various tendencies work their way through my system (or better: through the system that is me), a house results. The key thing here is that all of this happens blindly, without the subsequent motive tendencies being directed by or ordained to some end.[17]

I think that this amounts to as complete a package as is available for someone emphasizing the role of the mechanical sciences in Spinoza's thought.[18] We simply find ourselves thrust into doing certain things and thus judge them good. That our case is much more complex than, say, a bee's, and accompanied with sophisticated ideas and consciousness, does not alter the basic picture in any fundamental way; any action of an individual is a resultant of its constituents' inertial tendencies.

The inertial reading, at least as Carriero presents it, has considerable initial plausibility: it succeeds in painting an impressive overall picture of a mechanistic theory of action. However, closer scrutiny reveals that it is not unproblematic. First, Spinoza draws from the *conatus* proposition, 3p6, a consequence that is not easily explicable on the inertial interpretation. The decisive argumentative move behind 3p12, '[t]he Mind, as far as it can, strives to imagine those things that increase or aid the Body's power of acting', is that it follows *directly* from 3p6 that we strive to do whatever

[16] Ibid., p. 141. [17] Ibid., p. 142.

[18] For another reading that emphasizes efficient causes, but one that seems to me more vulnerable than Carriero's position to criticism that may be presented by the proponents of the teleological reading of Spinoza's *conatus* doctrine, see Parkinson 1981, p. 7.

increases our power of acting.[19] That this is what Spinoza has in mind is suggested even more strongly by the sibling proposition 3p13 and its demonstration, in which Spinoza claims that our mind does not just strive to continue thinking whatever ideas it has, but it strives to exclude bad ideas (by 3p9 which is, in essence, derived from 3p6), and can do this only with the help of good ones (by 2p17).[20] But it is hard to see how anything of this sort could be derived from the *conatus* principle interpreted as a principle stating that unless interfered with 'things will continue as they are'.[21] And since nothing that Carriero elsewhere says alleviates this trouble – nor can I see how anything available to him could – his interpretation seems to be in conflict with 3p12 and 3p13, and the way in which Spinoza demonstrates them.

Second, it is unclear why Spinoza would feel any need to argue for a generalized or metaphysical form of the principle of inertia in 3p6, because he has proven its physical variant already much earlier in the *Ethics*, and, what is more, with an argument revealingly different from the one located in 3p6d:

> [A] body in motion moves until it is determined by another body to rest; and [. . .] a body at rest also remains at rest until it is determined to motion by another.
>
> This is also known through itself. For when I suppose that body A, say, is at rest, and do not attend to any other body in motion, I can say nothing about body A except that it is at rest. If afterwards it happens that body A moves, that of course could not have come about from the fact that it was at rest. For from that nothing else could follow but that body A would be at rest.
>
> If, on the other hand, A is supposed to move, then as often as we attend only to A, we shall be able to affirm nothing concerning it except that it moves. If afterwards it happens that A is at rest, that of course also could not have come about from the motion it had. For from the motion nothing else could follow but that A would move. Therefore, it happens by a thing which was not in A, viz. by an external cause, by which [NS: the Body in motion, A] has been determined to rest. (2le3c)

It is very important to appreciate the fact that here Spinoza's point is an application of the thesis that, to borrow Leibniz's words once again, it is

19 Cf. Allison 1987, pp. 135–6, 235; Curley 1988, p. 115; Della Rocca 1996, pp. 213–15. It should be noted that in the early *Short Treatise* Spinoza expresses this idea in words that could not be clearer: 'For it is evident that no thing, through its own nature, could strive for its own destruction, but that on the contrary, each thing in itself has a striving to preserve itself in its state, and bring itself to a better one' (*KV* 1.5; C, p. 84; G 1, p. 40).

20 I say 'more strongly' because even if 3p12 were read as saying that when a mind happens to have a power-increasing thought, it just continues, as the inertial reading would have it, existing with that thought, there is nothing that would debar us from proving similarly that the mind must continue to exist *with any power-decreasing thought it happens to have*; and this, of course, is not something 3p13 would allow. So that proposition appears to go against the inertial reading more strongly than 3p12.

21 Carriero 2005, p. 134. Of course, this is a problem for any reading, teleological or not, according to which *conatus* is about preserving the prevailing state.

'a principle of metaphysical necessity' that 'each and every thing remains in its state [. . .] until something changes it' (AG, p. 172). Spinoza feels no need to refer to God's power, expression, or impossibility of self-destruction – as he does in the official demonstration of the *conatus* principle – to establish just that a simple body,[22] if not interfered with, retains its kinetic state for ever. Moreover, the above-quoted point is bound to be generalized, due to the parallelism thesis of 2p7, so that it applies to all things, no matter what their attribute is. Also 2p17,[23] as I read it, says that our mind continues to perceive an outer object as present until affected by something else, just as a mental principle of inertia would have it. So, having said all this already in the second part of the *Ethics*, it would seem gratuitous to once again undertake the task of proving that left to their own devices, things remain the way they are. At the very least, one would expect Spinoza to make a reference to the already secured 2le3c and possibly construct his *conatus* argument around it and the parallelism thesis.

Third, as is commonly agreed, the *conatus* doctrine includes the idea that things are able, through themselves, to *resist* destructive external causes. As far as I can see, the proponents of the inertial reading have a hard time saying where such resistance or true opposition could come from. This is understandable given that, to speak through Leibniz's mouth for the last time, it is one thing 'to retain a state until something changes it', and quite another thing to 'resist changing' (AG, p. 172). The latter is what Spinoza wants to prove in 3p6; but as I have argued above, this requires bringing into play dynamic notions that are not contained in the inertial account. Read in this way, Spinoza would thus be in no better a position than Descartes and Hobbes on the issue of resistance.

Fourth and more generally, the inertial reading seems to leave us very little to work with when faced with many metaphysical questions. Most importantly, problems pertaining to individuation seem difficult for it. Now, what makes a complex thing a unified individual persisting over time is, for Spinoza, obviously its *conatus*.[24] But if any complex thing is just, as Carriero puts it, 'the totality of its motive tendencies' that are, in turn, 'set by the motive tendencies of their components',[25] with nothing directing or shaping the individual as a whole, the individual's identity seems to become a dangerously fragile affair; why would the innumerable

[22] Judging from 2le3a2, 2le3 concerns simplest bodies.

[23] 2p17 reads: 'If the human Body is affected with a mode that involves the nature of an external body, the human Mind will regard the same external body as actually existing, or as present to it, until the Body is affected by an affect that excludes the existence or presence of that body.'

[24] Cf. Garrett 1999, p. 330; Barbone 2002, p. 100. [25] Carriero 2005, pp. 133–4.

motive tendencies out of which a complex thing is composed, left to their own devices, form anything capable of more than a very precarious and momentary existence? Indeed, does this not mean that any complex thing is demoted to being a mere aggregate, *unum per accidens*, falling short of having genuine unity and thus of being a true individual? It seems that the inertial line of thought, if taken as exhaustive of Spinoza's position concerning the being of finite individuals, leads to unwelcome results. To my mind, denying the importance of essences lurks behind this;[26] but I think it has already been shown how problematic this contention is in Spinoza's case. I admit that Spinoza probably does not have a well-formed theory of how the relationship between the striving of a complex individual and the strivings of its simpler parts should be conceived;[27] despite this, he obviously thinks that a complex thing's essence has, as does the thing's striving power, a particular character not simply derivable from its constituents' essences or powers. To my mind, the fact that the inertial reading does not appreciate this point results from its overall tendency to overemphasize the physical aspect and downplay the metaphysical aspect of Spinoza's thought.

So, its numerous merits notwithstanding, I do not think that the inertial reading – according to which Spinoza's views on *conatus* and human action are built on the idea of 'blind' (i.e. mechanistically conceived) efficient causes – is, in the end, the position to be endorsed. We must look for another interpretation of Spinoza's doctrine.

THE TELEOLOGICAL READING

Fairly recently, a considerable number of prominent scholars have claimed that, in spite of his comments to the contrary, Spinoza was not altogether hostile towards all kinds of teleology; and it has been argued that the *conatus* doctrine should be understood as teleological in character. Also the problems of the inertial interpretation raise the question as to whether some kind of teleological reading, despite swimming against the tide of tradition, would turn out to be the best way of interpreting Spinoza.

Before proceeding any further, it is necessary to outline, if only roughly, what is meant by 'teleology'. There are, of course, several possible ways to approach this question; the issue is as complex and fine-grained as it is important for understanding Spinoza's conception of agency. We can begin

[26] See Carriero 1995 (p. 272) for an argument for the diminished role of essences in the post-Aristotelian world.

[27] See especially *Ep*30; but also *Ep*32 (S, p. 849; G IV, p. 171) and 2p24d.

by noting that it has become customary in Spinoza scholarship to speak about teleology of three different kinds: *unthoughtful*, *thoughtful*, and *divine* teleology. The first concerns non-cognitive beings such as rocks and trees and does not involve conscious thought, the second concerns cognitive finite beings that have thoughts (most notably of future states of affairs), and the third concerns God. Scholars are often explicit about how they stand with regard to these: one may say, for instance, that Spinoza denies any finality with regard to God but that the actions of cognitive agents are end-governed, i.e. that there is thoughtful teleology in Spinoza.

It seems to me that the distinction, vital in all philosophy of science, between explanatory models and ontological frameworks is important to keep in mind here: in Spinoza studies, as elsewhere nowadays, by 'teleology' is most often meant a certain *scheme of explanation* in which items are explained by citing their (beneficial) future effects. As Andrew Woodfield writes, the standard way of explaining teleologically 'why an event occurred is to say that it occurred *in order that* a second event should occur, or *in order to* produce a certain result'.[28] For instance, one might claim that a black bear has sharp front teeth and flat back teeth because this allows tearing and chewing food efficiently, which, in turn, occurs in order that the bear survives.[29] But this should be distinguished from one's stand in ontological matters. If someone claims purposes or final causes to be fundamental for the causal architecture of the world so that things operate according to them – for instance, if one says that the heart *is structured with an end in view*, namely circulating blood, and that is why it exists and operates the way it does – a particular teleological metaphysic is thereby being put forward. The most influential teleological ontological framework in the history of thought has, of course, been that developed by the Aristotelian tradition, and we should keep in mind that Spinoza's intellectual milieu was to a great extent shaped by its late scholastic variant.

Explanatory schemes and ontological frameworks are deeply intertwined; I take it as uncontroversial that our ontological commitments largely determine, in ways we often do not realize, the types of explanations we are willing to accept. Should one be inclined to champion an ontology

[28] Woodfield 1976, p. 16. Slightly later on the same page, he elaborates:

> Anyone who asserts that a sequence of events was teleologically directed to the end at which it stopped takes himself to be making a substantive claim which goes beyond the claim that the sequence did in fact end at that point. He is claiming not only that the earlier parts led to the end, but that there was a press of events in that direction, such that the later event provides an understanding of why the earlier events occurred.

[29] The example is lifted, in a slightly modified form, from Garrett 1999, p. 310.

similar to that of Aristotle, teleological explanations would appear completely adequate, in fact unsurpassable, as they would be regarded as referring to and being based on how things are. But should one hold that in the final analysis reality contains only, as the science of mechanics assumes, lumps of moving matter that transfer their motion via impacts to each other, teleological explanations would appear to lack a proper ontological grounding; as a consequence, any talk about ends is most likely bound to appear suspect, at best reducible into other more adequate types of explanations involving only non-teleological concepts. Evidently, the success of the theory of evolution underlies the fact that most contemporary philosophers consider teleological explanations in natural sciences to be reducible to explanations citing only efficient causes – for instance, of the bear example one might say that there has been an evolutionary process in the course of which only those bears with a certain kind of dental arrangement have been able to procreate and increase their numbers – but there are still also those who think that at least biology contains ineliminably teleological aspects.

Proponents of the inertial reading see, as we have observed, Spinoza as rejecting the Peripatetic ontology in its entirety and, consistently with this, as being hostile towards all teleological explanations. The position of the scholars assigning teleology to Spinoza does not, however, seem to be as clear-cut. For instance Garrett speaks mostly about teleological explanations and offers us a definition of teleology – which I take to be a very important articulation of the general understanding of teleology in recent Spinoza scholarship – as 'the phenomenon of states of affairs having etiologies that implicate, in an explanatory way, likely or presumptive consequences of those states of affairs'.[30] However, as this says nothing specific about ontological matters, it is not entirely clear what exactly it means to find a thinker a 'teleologist' by these standards. In any case, I think it is most useful to keep these preliminary observations in mind when examining Spinoza's stand on 'teleology' – a topic much of this chapter focuses on.

It should be noted at the outset that apparently no commentator to date has held that Spinoza would have accepted divine teleology in any form. This is understandable, for much of the appendix of the opening part of the *Ethics* is explicitly dedicated to showing as absurd the idea, common in Spinoza's time, that God's actions are end-directed. According to Spinoza, the root of all false beliefs concerning God's causality is the fact

[30] Garrett 1999, p. 310.

that men commonly suppose that all natural things act, as men do, on account of an end; indeed, they maintain as certain that God himself directs all things to some certain end, for they say that God has made all things for man, and man that he might worship God. (1app; C, pp. 439–40; G II, p. 78)

However, it is better to postpone the detailed examination of the argument Spinoza puts forward here, because it has a bearing on the issue of unthoughtful and thoughtful teleology as well.[31]

Of the teleological interpretations offered, Don Garrett's 'Teleology in Spinoza and Early Modern Rationalism' is not only probably the most detailed one but also one that discusses at length the nature of the *conatus* doctrine. Garrett's reason for regarding the *conatus* doctrine as teleological is rather straightforward:

Spinoza seems to hold that each thing has at least some causal power whose exertion is a striving or tendency of the thing to persevere in being. This doctrine provides an obvious avenue for explaining the behavior of singular things by appeal to the self-preserving tendency of that behavior.[32]

Some pages later he explicates the way in which Spinoza might be thinking:

The teleological functioning of an 'end' or 'final cause' arguably requires at least that something be *selected from alternatives* in a way that essentially involves some kind of *goodness or fitness of its likely or presumptive consequences* relative to those of other alternatives. For the actions of singular things, including human beings, the idea of such selection makes sense. [. . .] Since singular things naturally pursue their own self-preservation – according to the *conatus* doctrine of E 3P6 – we can appeal to a specified action's (likely or presumptive) beneficial consequences in order to explain why the singular thing 'selected' *that* course of action over the alternatives that were equally possible relative to the power of the surrounding objects.[33]

So, because Garrett defines teleology in terms of an explanatory framework, and his interpretation of 3p6 allows us to explain the fact that x does A by indicating A's consequences, the formulation in question qualifies as teleological. This, in turn, would mean that a type of teleology would apply to all finite things, also making non-cognitive entities subject to teleological selection processes 'capable

[31] A terminological note is in order at this point. Curley (1988, p. 107; 1990, p. 48) rightly observes that although the term *conatus* might be translated – apart from 'striving' or 'endeavour' – as 'trying' or 'making an effort', which suggest conscious goal-oriented behaviour, we should keep in mind that in Descartes's philosophy, *conatus* has a technical use (by which Spinoza was influenced) not implying any kind of thinking or wanting; cf. Della Rocca 2008b, p. 145. In other words, Spinoza's terminological choice should not tempt us to regard the doctrine as *prima facie* teleological (here I would disagree with Bennett 1984, p. 245).

[32] Garrett 1999, p. 313.

[33] Ibid., p. 316.

of selecting and producing states of affairs on the basis of their typical or presumptive consequences'.[34]

Curley and Nadler also put forward teleological interpretations of 3p6. Curley interprets it as having the form *If A would help x, x will do it* and as being quite general in character, intended to apply also 'to things which are not capable of having thoughts about the future'.[35] In the same vein as Curley and Garrett, Nadler thinks it best to interpret 3p6 as putting forward a general form of teleology concerning finite things: 'Thus, *all* individuals have a basic kind of teleological behaviour, in so far as they strive to do what best preserves their being.'[36]

Obviously, an important reason for interpreting the *conatus* doctrine teleologically stems from the fact that after, and on grounds of, 3p12 (which is ultimately derived from 3p6) Spinoza makes claims that, as Garrett puts it, seem 'intended to license teleological predictions and explanations of human actions'.[37] For instance 3p28, '[w]e strive to further the occurrence of whatever we imagine will lead to Joy, and to avert or destroy what we imagine is contrary to it, *or* will lead to Sadness', has been singled out as such a claim.[38] Now, Bennett interprets 3p6 non-teleologically and claims thereby the following problem to arise. The best argument we can devise for the *conatus* principle is, according to Bennett, as follows: from 3p4's *If x does A, then the doing of A does not destroy x* it can be inferred that *If x does A, then the doing of A does not tend towards x's destruction*; and with some stretching we can arrive from this at *If x does A, then the doing of A tends towards x's preservation.* However, the main trouble with this argument and its non-teleological conclusion is that it cannot support positive predictions of behaviour, for it only tells that x will *not* do something because that will not help it. But as was just pointed out, the deductive progeny of 3p6 seems to be teleological in nature; hence, Bennett holds Spinoza to illegitimately strengthen the *If x does A, then the doing of A tends towards x's preservation*, which would be more or less

[34] Garrett 1999, p. 325; for the claim that teleology pertains to all Spinozistic individuals, see esp. p. 330.

[35] Curley 1988, p. 164. Curley thinks, following here Bennett (1984, p. 245), that the phrase *quantum in se est* speaks for understanding 3p6 teleologically; but based on the discussion in Ch. 3 above, I do not think this to be the case. Moreover, Curley makes the important claim that Spinoza takes from the general *If A would help x, x will do it* of 3p6 to follow the 'teleological-*cognitive*' doctrine of e.g. 3p28 that has the form *If x* thinks *A would help it, x will do it.* Although Curley has a rather positive view of the derivation of 3p6 and the teleological doctrine he finds in it, he does not think the move from 3p6 to the cognition-involving 3p28 to be licit.

[36] Nadler 2006, p. 199. [37] Garrett 1999, p. 314.

[38] According to Bennett (1984, p. 245), from 3p12 onwards there are eleven such propositions in the third part of the *Ethics*, all derived from 3p12 and 3p13; see also Garrett 1999, p. 314; Lin 2006, pp. 320–1.

available for him, to the teleological and much stronger *If A is helpful to x, x will do it*, which has behaviour in the consequent and can thus, unlike the non-teleological version, be used in explaining human action.[39]

As just noted, proponents of the teleological reading are also eager to emphasize the apparently teleological use Spinoza makes of 3p6 in the rest of the *Ethics*. But Garrett thinks the derivation of 3p6 to be defendable and the doctrine it presents teleological in the first place, so his Spinoza would be innocent of any malpractice.[40] Although I think that this line of interpretation overemphasizes the teleological aspects in Spinoza's theory of human action, the fact that especially the third part of the *Ethics* does contain material that can quite naturally be seen to allow explaining things teleologically is to be taken seriously when interpreting Spinoza's philosophy in general, and the *conatus* doctrine in particular.

Whether or not the *conatus* doctrine is regarded as licensing teleological explanations has major repercussions for the way in which Spinoza's theory of human action and motivation is understood. Much hinges on how one interprets a passage we have already encountered while discussing the inertial reading:

> What is called a final cause is nothing but a human appetite [. . .] For example, when we say that habitation was the final cause of this or that house, surely we understand nothing but that a man, because he imagined the conveniences of domestic life, had an appetite to build a house. So habitation, insofar as it is considered as a final cause, is nothing more than this singular appetite. It is really an efficient cause, which is considered as a first cause, because men are commonly ignorant of the causes of their appetites. (4pr; C, p. 544; G II, p. 207)

Curley comments on that passage as follows:

> Spinoza has a way of making talk of final causes acceptable, but it is precisely the way Bennett says he would reject: when it appears that something is being explained by a subsequent result, what is really happening, if the explanation is legitimate at all, is that we are explaining a human action by appealing to the person's anticipation of the consequences to be expected from that action, his desire for those consequences, and his resultant desire to perform the action.[41]

Jarrett similarly defends the view that Spinoza espouses *thoughtful* teleology and sees no problem in him accepting claims of the form, *If x thinks A will*

39 Bennett 1984, pp. 243–5; see also 2001, pp. 221–2.

40 Lin (2006) defends Garrett's interpretation by arguing, to make the case against Bennett (see n. 12 above), that the causal powers of ideas are partly individuated by external causes and that Spinoza thus accepts, and is entitled to, teleological explanations.

41 Curley 1990, p. 48.

help x, x will endeavour to do A.[42] Now, I find this to be quite a natural way to read the house-building passage, if we focus on it alone; but I also think that when that passage is considered in connection with what I take as the accurate interpretation of the nature of the *conatus* doctrine, something still more intricate seems to be going on in it. As it is a truly revealing and important piece of text, I will present my own understanding of it in the next section.

Garrett presents an especially noteworthy interpretation of the Spinozistic theory of human motivation. He notes the talk about appetite in 3p9s to be 'an application to human beings of the more general doctrine of 3p7' that the *conatus* is the actual essence of things, 'from which', Garrett holds, 'the various actual properties and actions of the human being follow'.[43] Now all this is, on the whole, correct. But Garrett wishes to establish a much stronger connection between actual essences and teleology. To my mind rather unexpectedly, he argues from the uncontroversial fact that, for Spinoza, an actual essence equals striving to persevere in being to follow that it 'constitutes, in itself, a general teleological selection process'.[44] It is precisely this final step that seems to me problematic. I will shortly present what I mean by this, but meanwhile we should note Garrett's reasons for his contention. Obviously, it is based on a corollary of his already cited teleological rendering of the *conatus* doctrine, 'we can appeal to a specified action's (likely or presumptive) beneficial [i.e. self-preservative] consequences in order to explain why the singular thing "selected" *that* course of action over the alternatives that were equally possible relative to the power of the surrounding objects';[45] and since Spinoza certainly holds actual essences to be the centres of our self-preservatory activities, Garrett thinks we can single them out as the loci of teleological selection processes. The crucial general principle concerning action explanations would then seem to be of the form, *If A is likely to promote self-preservation better than any other action within x's power, x endeavours to do A.*

There are, however, serious problems in interpreting *conatus* as a teleological principle and Spinoza as a committed teleologist. Before offering a more thorough exposition of why this is the case, I would like to make three brief critical observations. First, an admittedly non-specific but still important point:

[42] Jarrett 1999, pp. 5–7, 10–12, 20–1. '[F]inal causes, understood as actual motivational states of the agent or object, do exist, and are the *efficient causes* of actions and events' (Jarrett 1999, p. 7). Della Rocca (2008b, p. 82) also holds a similar position.

[43] Garrett 1999, pp. 322–3. [44] Ibid., p. 327. [45] Ibid., p. 316.

the *Ethics* simply gives a strong general impression of being anti- or non-teleological by design. In other words, I think that if there is a *prima facie* case for anything in Spinoza's thought, it is against all teleology; this is also witnessed by his traditional reputation as an anti-teleologist. All this is, of course, still only tentative, but nevertheless enough for me to think there is much truth in Bennett's overall assessment of Spinoza's action-theoretical tendencies: '[M]any aspects of Part 3 cannot be understood unless one grasps that Spinoza is trying to develop a nonteleological theory of human motivation.'[46]

Second, the problems plaguing the inertial account notwithstanding, the point that Spinoza is trying to carry out a reductionist programme according to which talk of ends can and should be translated into non-teleological language appears well-founded. I think that Bennett is exactly right in holding that 'Spinoza retained the *language* of thoughtful teleology because he thought he could show that, properly understood, it is not teleological after all.'[47] For instance, '[w]hat is called a final cause is nothing but a human appetite insofar as it is considered as a principle, *or* primary cause, of some thing' (4pr; C, p. 544; G II, p. 207) lends credence to Bennett's claim.[48] Moreover, it should be noted that this piece of text contains an expression that has, with reason, been claimed to support the anti-teleological cause:

> The phrase 'nothing but' is the battle-cry of the reductionist; it heralds an attempt to get teleology out of the picture. If Spinoza had no such purpose [. . .], this performance of his would be pointless. Why say appetite is 'nothing but' [. . .] and then follow that with something that he thinks has nested within it the whole normal teleological story?[49]

Why indeed? I find this very convincing. As a consequence, I think Carriero is on to something important when he contends, 'if we were not ignorant of the causes of our appetites [. . .] we would drop talk of final causes and stick with appetites or efficient causes. So while it is true that Spinoza says final causes are appetites, I think we need to be alive to a hint of an error theory here.'[50]

Third, 3p6 echoes the seventeenth-century conservation laws of motion, and Spinoza probably saw just as little teleology to be involved in his *conatus* principle as is involved in those laws.[51] Rice articulates this point nicely with

[46] Bennett 1984, p. 215; see also Parkinson 1981, p. 6; Bennett 2001, pp. 209–10, 217.
[47] Bennett 2001, p. 211; see also 1990, p. 53.
[48] See also 1app (C, p. 442; G II, p. 80), 3p2s (C, p. 497; G II, p. 143), and 3p9s.
[49] Bennett 2001, p. 217. See also Carriero 2005, p. 141.
[50] Carriero 2005, p. 141.
[51] On this, see n. 5 above.

regard to mental individuals: '[T]he primary model which Spinoza uses for his account of mental *conatus* is simply lifted from the physics of his time, [. . .] there is no reason to assume that such a model would require appeal to teleological danglers.'[52] In other words, a teleological rendering of the *conatus* principle is uneasily reconcilable with Spinoza's mechanistic tendencies.

The greatest difficulty, however, with the teleological reading is that it does not square with Spinoza's basic ontological commitments. Garrett's point is that inasmuch as finite modifications, especially human beings, do the things they do because of the salutary effects those doings have, it follows that the *conatus* doctrine licenses explanations of the teleological kind. However, if we take a closer look at the causal architecture Spinoza develops, the problems pertaining to reading Spinoza from this angle begin to take shape. In his system, things are what they are because their essences follow, with geometrical necessity, from God's nature (1p16, 1p17s). Now, as I have already noted, no scholar I know of is willing to claim that Spinoza's God would have anything to do with teleology. This, in turn, is based on the fact that 1app contains a number of arguments directed against (at least) divine teleology. It is of the utmost importance to have a clear grasp of what is going on in that appendix, a topic I will address below, but for now it suffices to recall it to be uncontroversial that God's causality, or the 'following from his nature', does not involve any finality.

Thus far we have established that things have the kind of essences they do because those essences follow from God's essence and that this following is non-teleological in character. Now, the individual essences determine how things act (3d2, 3p7d) and, to a large extent, what they passively undergo or do (2p16, 3d2). In other words, the fundamental ground for things' existence and manner of operation lies in their essences – and from each one of them, Spinoza holds, some effects follow (1p36). There is nothing in all this, I believe, that Garrett would not be willing to accept; after all, he emphasizes the idea of essences as centres of causal activity. Clearly, what he wants to do, in addition to this, is to establish a connection between Spinoza's essentialism and teleology. Of course, these two do not as such contradict each other (were they to do that, all kinds of Aristotelianism would be doomed from the start); but the connection becomes strained in Spinoza's geometry-inspired framework. As far as I can see, nothing Spinoza says suggests that he would see the 'following' from finite things' essences to be teleological any more than it is in God's case.

[52] Rice 1985, p. 249.

At this point we should take a new look at the argument located in 1app which Spinoza himself clearly regards as the principal one against teleology. Spinoza claims to have 'sufficiently established' that 'Nature has no end set before it' and 'all final causes are nothing but human fictions' by such propositions as 1p16, 1p32c1, 1p32c2, 'and all those [propositions] by which I have shown that all things proceed by a certain eternal necessity of nature, and with the greatest perfection'. The specifically mentioned passages read as follows:

> From the necessity of the divine nature there must follow infinitely many things in infinitely many modes (i.e., everything which can fall under an infinite intellect.) (1p16)

> From this it follows, first, that God does not produce any effect by freedom of the will. (1p32c1)

> It follows, secondly, that will and intellect are related to God's nature as motion and rest are, and as are absolutely all natural things, which (by p29) must be determined by God to exist and produce an effect in a certain way. For the will, like all other things, requires a cause by which it is determined to exist and produce an effect in a certain way. And although from a given will, *or* intellect infinitely many things may follow, God still cannot be said, on that account, to act from freedom of the will, any more than he can be said to act from freedom of motion and rest on account of those things that follow from motion and rest (for infinitely many things also follow from motion and rest). So will does not pertain to God's nature any more than do the other natural things, but is related to him in the same way as motion and rest, and all the other things which, as we have shown, follow from the necessity of the divine nature and are determined by it to exist and produce an effect [*operandum*] in a certain way. (1p32c2)

It is especially noteworthy that in these passages, Spinoza keeps invoking his monistic framework. But he does not do this, as has been suggested,[53] because he is aiming at denying divine teleology only; he does it to remind us that from a metaphysically adequate viewpoint, there is only one causal agent, God, from whose nature everything follows with necessity that excludes choice or free will. For determining *the scope* of Spinoza's denial of teleology the end of 1p32c2, 'things which [. . .] follow from the necessity of the divine nature and *are determined by it to exist and produce an effect* in a certain way',[54] is particularly revealing: since it also claims the *causation of finite things* to follow 'from the necessity of the divine nature', God's geometrical necessity obviously pertains to their causal activities, too – which is exactly what we should expect given Spinoza's geometry of

[53] See especially Curley 1990, pp. 40–1; but also Garrett 1999, p. 315; Lin 2006, p. 323.
[54] Emphasis added.

power. As a consequence, the main argument in 1app seems to be, after all, directed against *all* finality: both in the case of infinite and finite causation, the necessity with which things take place rules out finality.

The aforesaid implies that Spinoza is not trying to derive a denial of teleology merely from causal determinism and the denial of freedom it entails, but from geometrical necessitarianism. However, as scholars have been quick to point out, it is a mistake to think that teleology would require causal indeterminism.[55] Things may very well be necessitated to act teleologically. What, then, has *any* kind of necessitarianism to do with getting teleology out of the picture?

Here we should be especially alive to the historical context in which Spinoza develops his philosophy and note with Woodfield that

> Renaissance scientists who rejected final causes were reacting against a concept which lay at the heart of Aristotelianism. It is difficult for us, conditioned as we are by their success, to imagine how things looked before the scientific revolution.[56]

This also applies to interpreting Spinoza: the teleology he is opposed to – *probably the only kind imaginable at his time* – is of the kind embedded in the Aristotelian framework. I think it is correct to say in general that, in that framework,[57] there was a strong tendency to view the final cause as *prior* to – perhaps we could even say *a constituent of* – everything else in the causal architecture of the world: it is the cause of causality and all other causes, also of the efficient cause, because the most fundamental feature of any thing's ontological make-up is the end for the sake of which it acts, the end it strives to actualize.[58] So it is correct to say that agents were seen, in one way or the other, as 'end-directed or end-governed', with powers that 'are organized around ends'.[59]

The outlined general stance persisted from Aristotle to the late scholastics, but underwent some considerable changes so that as time went by, final causality came increasingly to be seen as pertaining to cognitive agents only.[60] This happened mainly because it was hard to see how any final cause could operate other than by setting the will into action: the end is something good in itself that, when contemplated, draws the will towards itself, makes the will desire the good; and then the will, as the efficient cause, brings about the intended end (this way the end can be temporally prior to

55 See e.g. Bennett 1984, p. 216; Curley 1990, p. 43; Garrett 1999, pp. 315–16.

56 Woodfield 1976, p. 4.

57 See also Ch. 2 above.

58 For an illuminating discussion of this, see Carriero 2005, pp. 107–20.

59 Ibid., pp. 112–13, 109. See also Des Chene 1996, pp. 180–1.

60 The account of this paragraph draws largely from Des Chene 1996 (esp. pp. 169, 189–94, 235, 248) and Pasnau 2001 (esp. pp. 305–9).

its effects). But this kind of purposiveness, or the proper operation of the final cause, clearly cannot apply to natural agents lacking intellect and will, and so their tending towards certain ends was regarded as something impressed on them by what was regarded by the scholastics as the rational agent *par excellence*, the Christian God.[61] In other words, non-rational beings act for ends, but only insofar as they are related to God.[62] The finality of natural agents is explicated through a model similar to that in terms of which human finality is understood, the purposive agent being the transcendent designer-God.

Given Spinoza's intellectual background, it seems evident that he is not interested in rejecting what for us is first and foremost *a particular style of explanation* – that it is in fact difficult, as Carriero notes,[63] to say to what extent our conception of teleological explanations would have been *even recognized* by thinkers of Spinoza's era is, I think, a thrilling display of the historical distance between us and them – but *aims at overthrowing a whole metaphysical tradition and replacing its ontology with a new, adequate one.* Recall the essentialist model: there is only one thing, modelled on geometrical objects, that exercises its power according to its definable essence. And so, just as in geometry, everything pertaining to God can be inferred from God's definition; and just as in geometry, the definition and everything deducible from it do not involve ends, purposes, or intentions: Spinoza explicitly endorses the traditional contention that final causes are excluded from mathematics concerned 'not with ends, but only with the essences and properties of figures' (1app; C, p. 441; G II, p. 79). As Mancosu explains, 'final causes had, in fact, traditionally been excluded from the realm of mathematical demonstrations'.[64] I think that Spinoza is attacking all kinds of Aristotelian doctrines of end-governedness, but his geometrical essentialism contrasts particularly sharply with the above-outlined late scholastic framework: geometrical entities cannot be end-guided already for the reason that they are not rational agents; and if the divine causality is like that pertaining to geometry, there is no way it, or anything it brings about, can involve purposiveness.

[61] This means that for the medievals but not for Aristotle, final causes cannot be explanatory without what Pasnau (2001, p. 307) calls 'intentional salience'. According to Pasnau's very helpful account, this kind of change in the notion of final cause took place largely because of Avicenna's influence.

[62] Pasnau (2001, pp. 308–9) estimates it as 'the standard medieval view' that 'final causality is possible only in virtue of a mind that grasps the end in question. So if nature does not act according to the divine mind, then there is no genuine acting for ends in nature, and hence no genuine teleology.'

[63] Carriero 2005, p. 106. [64] Mancosu 1996, p. 17.

However, it is hard to deny that the *Ethics* contains propositions licensing teleological explanations in their modern sense. If any claim that explains an item by its presumptive consequences qualifies as teleological, how could, for example, '[w]e strive to further the occurrence of whatever we imagine will lead to Joy' (3p28) not allow explaining an action by its supposedly joyful consequences? Material enabling teleological explanations thus seems to creep into Spinoza's system, after all. What could he say about that, or we on his behalf?

The first thing to note is that even if some of Spinoza's propositions licensed teleological explanations, a genuinely teleological ontology would still not be forced on him. Of course, Garrett only claims that an idea of a teleological selection process 'capable of selecting and producing states of affairs on the basis of their typical or presumptive consequences'[65] can be combined with Spinoza's view of essences as entities that 'must causally explain, rather than be causally explained by, such other properties of the person as judgments about the good'.[66] It is, however, difficult to see how Spinoza could think his essentialism to be combinable even with this conception of teleology; I suspect that he would regard any kind of 'teleological selection process' having an effect on our actions as illegitimately threatening the ontological priority of essences. By postulating such a process we risk lending ourselves to criticism that Spinoza presents by saying, '[w]hatever conduces to health and the worship of God, they have called *good*; but what is contrary to these, *evil*' (1app; C, p. 444; G II, p. 81). Spinoza's own stand, obviously against this, is that something is good because we desire it (3p9s), desire being our essence itself.[67] For the present purposes it suffices to note that there is in Spinoza a clear insistence on the pre-eminence of essences: there are no goods independent of our essences, for our essence-originating activities determine what is good. Consider once again how sharply this position contrasts with the above presented traditional view according to which things are structured around their ends; this sort of ontological priority of ends, making essences dependent on their ends, is definitely something Spinoza wants to avoid. On my view, which I will present more thoroughly below, an auxiliary argument he puts forward in 1app (C, p. 442; G II, p. 80) is meant to be an undisguised statement to this effect: the doctrine of ends reverses the order of nature, '[f]or what is really a cause' and 'by nature prior', i.e. the essence of a thing, 'it considers as an effect, and conversely' – that is, as if the essence would be what it is because of an end, or because of what it causes.

[65] Garrett 1999, p. 325. [66] Ibid., p. 323. [67] See e.g. 3p56d.

To conclude, I can only deem the teleological reading, because of its aforementioned problems, undersupported. And so it seems, rather surprisingly, that despite the stimulating discussion in which so many prominent scholars have taken part, we still do not have to hand an adequate interpretation of the meaning of the *conatus* doctrine and of the Spinozistic theory of human agency. In what follows, I will attempt to remedy this situation. Given the kind of ontology I have been outlining, I think it is well-grounded to claim that Spinoza adopts, as Bennett and Carriero emphatically argue, a reductionist position concerning teleological terminology; so just as Spinoza holds final causes to be 'nothing but' non-teleological appetites, he is probably thinking that all apparently teleology-involving passages can be converted into non-teleological explanations that are based on a true ontology containing 'nothing but' harmless non-teleological elements. It is another question whether he really can pull off such a reductionist programme; but the conviction that this can be done most certainly seems to be there.[68] Although the inertial rendering of this programme is not, to my mind, satisfactory, I will argue in the next section that there is another kind of non-teleological position available to Spinoza, unearthable from his basic ontology.

CONATUS AS THE PRINCIPLE OF PERFECT ESSENCE REALIZATION

The guiding idea of my interpretation is that Spinoza elaborates, especially in the latter half of the opening part of the *Ethics*, a particular metaphysic of essences and their powers that is designed to capture the basic causal architecture of the world. Most importantly, the doctrine I have dubbed 'geometry of power' shapes the *conatus* principle, which thus cannot be properly grasped apart from its metaphysical moorings. It is especially important to recall the connection Spinoza's views on causation have to geometry and the conception of formal causation involved in it; this model of internal causation pertaining to geometrical things underlies Spinoza's attempts to construct a theory of finite things' action which, despite having nothing to do with finality, allows him to claim things to be endowed with something that directs the way in which their causal powers are exercised. Thus, essences also provide in the temporal realm the required ontological foothold through which the world is ordered, salvaging the physical reality of the new science from being a chaotic flux. More substantially, as already

[68] For a similar estimation, see Rice 1985, p. 249.

my analysis of the proof of 3p6 suggested, things are endowed with *conatus*, that is, *striving to drive themselves through opposition to produce effects determined by their definable essence alone*. The fact that things are endowed with power (and are so because they express God's power) grounds the thesis that they genuinely *resist* any opposition, not just (indifferently) maintain whatever prevailing state they may have as long as they are not interfered with.

To frame the issue under scrutiny, we can examine a constellation of interrelated metaphysical concepts located in the pages of the *Ethics*, beginning by taking a look at that of *determination*. Now, this concept occupies what may be regarded as a general or even neutral position in Spinoza's system, for both eternal and temporal existents can be said to determine each other. It is only *sub specie durationis* that determinations equal helps and hindrances, cooperation and competition. Here, however, we do not need to focus on these differences but note that things can be said to be *externally* and *internally* determined. Sometimes Spinoza writes as if finite things were exclusively externally determined;[69] but many other passages imply that this is not his complete view. 2p29s expressly states that things are determined internally as well;[70] moreover, as also this survey of the related concepts will show, were this not so, many aspects of his ethical project would remain incomprehensible. So obviously such propositions as the famous 1p28 consider finite things *insofar as* they are determined externally, from the 'common order of nature', which, however, does not prevent them from being also, at least to some extent, internally determined, that is from existing and acting as determined by one's own nature alone.

Moreover, to the extent things are internally determined they are, for Spinoza, *free*. Recall,

> [t]hat thing is called free which exists from the necessity of its nature alone, and is *determined to act by itself alone*. But a thing is called necessary, or rather compelled, which is *determined by another* to exist and to produce an effect in a certain and determinate manner. (1d7, emphasis added)

And since freedom and self-determination obviously equal what Spinoza sometimes calls bringing about effects that can be understood from or through the laws of one's nature alone,[71] things can be said to be *autonomous*

[69] See e.g. 1p28; *Ep*58 (S, p. 909; G IV, p. 266).

[70] See also defaff1; 4p37s2; *CM* II.4 (C, p. 322; G I, p. 256); *TTP* III (S, p. 417; G III, p. 46). For a related discussion, see Schmidt 2009, pp. 302–3.

[71] 3p2s, 4d8, 4p2d, 4p18s, 4p19, 4p24, 4p35.

to the extent their prevailing state is determined only by themselves. Further, the definition of action and passion, 3d2, makes it clear that this is coextensive with being *active*;[72] and as we have seen, in dynamic terms this equals having a high level of *power of acting*; in ethical terms, being *virtuous*. To close the list, since 'the perfection of things is to be judged solely from their nature and power' (1app; C, p. 446; G II, p. 83), the obvious outgrowth of all this is that if my interpretation is correct, *conatus* is not just a principle of persevering in existence but a principle of persevering in *perfect existence*,[73] of striving to exist in an autonomous, free, virtuous, active, and self-determined state in which is instantiated not only the striver's essence but also everything that follows from it alone.[74] To the extent any thing succeeds in not only existing but also in being self-determined, it is endowed with power of acting. In light of the first two chapters of the present work, this means that – and here we should appreciate the beautiful unity of Spinoza's thinking – a variant of the practically universally recognized *essence/property ontology is the driving force behind this scheme of things.*

So on my interpretation, things exercise their power not only to exist but to exist according to their definable essences alone; or, they strive to bring about being determined by the unhindered realization of their essences. Thus, I would call *conatus* the principle of perfect essence realization,[75] and its central idea can be stated as follows: *each and every singular thing is a powerful entity which, when it encounters opposition, strives to exist and to*

[72] Judging from what Vere Chappell (2005, p. 127) writes about self-determination and agency in seventeenth-century philosophy, Spinoza is a veritable representative of his times in making these connections.

[73] Despite commenting at one point, '[p]erfection and imperfection [. . .] are only modes of thinking' (4pr; C, p. 545; G II, p. 207), there is a sense of perfection and imperfection Spinoza regards as philosophically approvable, that which is grounded in the changes of our power of acting: '[W]hen I say that someone passes from a lesser to a greater perfection, and the opposite, [. . .] we conceive that his power of acting [. . .] is increased or diminished' (4pr; C, pp. 545–6; G II, pp. 208–9). Moreover, compare 3p11s with 3p15d; other relevant passages include defaff1, 4p18s; *Ep*19 (S, p. 808; G IV, p. 89). See also Spindler 2005, p. 43; Lin 2006, p. 326; Della Rocca 2008b, pp. 85–6; Garrett 2008, p. 13; LeBuffe 2010, pp. 143–4.

[74] See also 4d8, 4p8d. As a matter of fact, it has become quite common to view *conatus* as a principle of power-enhancement; however, I have not been able to locate a convincing argument for what would warrant this. For a discussion on a notable type of interpretative stand, see below.

[75] Although this way of interpreting the *conatus* principle may seem rather controversial, its general thrust agrees, I think, with some readings hailing from the French tradition (see e.g. Matheron 1988, p. 49; Jaquet 2004, p. 64; Spindler 2005, p. 27). I sympathize most with what Victor Delbos (1950, pp. 122–3, translation mine) holds; according to him, *conatus* is to be 'conceived as the realization of an essence [. . .] Spinoza's determinism is geometrical rather than mechanical [. . .]: individuality, with the striving pertaining to it, [. . .] is a singular definition that realizes itself.' See also Schmidt 2009, pp. 301–4.

bring about things derivable from its own definition alone. I will elaborate on this further when I discuss Spinoza's views on human agency; for now it suffices to note that this licenses positive predictions of behaviour without being at least overtly teleological, and certainly not teleological in any Aristotelian sense.

The present interpretation explains the puzzling fact we have already encountered, namely that in 3p12d and 3p13d Spinoza claims it to follow from 3p6 (at least via 3p9) that things strive to *increase* their power of acting: given that 3p12 and 3p13 concern situations in which we are affected by external causes and thus not in a purely autonomous state, in such situations *conatus* as interpreted here amounts to perfection-increasing, not merely to preserving the already attained state. An important line of interpretation concerning these matters is as follows. According to Allison, things always struggle with external causes, and so they must 'constantly endeavor to increase their power, merely to preserve their actual level of existence';[76] in a similar vein, Curley contends that the idea of 'increased power of action' follows from the fact 'that powerful external forces are ranged against us'.[77] Della Rocca offers what I regard as the most elaborate variant of this type of approach and suggests that the striving for self-preservation leads to striving for more power, because we anticipate that the future comes with many threats to our existence, and 'it is in our own interest to accumulate as much power as possible to be able to meet these various threats when they occur'.[78] But the propositions under scrutiny do not mention the future; moreover and more generally, it does not seem to follow from the idea of striving to persevere in the present state that things strive to increase their own power just to be on the safe side of survival as it were. If prolonged existence is all one is striving for, maintaining the status quo is enough if it means living under a predictable (or 'safe') set of non-lethal passions that allows a stable existence, which is something 3p12 and 3p13 deny. But if *conatus* is understood along the lines I suggest, no non-optimal state is enough and leads to striving to increase one's power or perfection – which may happen by striving to continue thinking good, or power-increasing,

[76] Allison 1987, p. 235; see also pp. 135–6. [77] Curley 1988, p. 115; see also p. 166.

[78] Della Rocca 2008b, p. 172; see also p. 156. The scholars defending this type of interpretation draw attention to *Leviathan* I.11. Hobbes states:

> So that in the first place, I put for a general inclination of all mankind, a perpetual and restless desire of power after power, that ceaseth only in death. And the cause of this, is not always that a man hopes for a more intensive delight, than he has already attained to; or that he cannot be content with a moderate power: but because he cannot assure the power and means to live well, which he hath present, without the acquisition of more. (*EW* III, pp. 85–6)

thoughts (3p12) as well as by striving to get rid of bad, or power-decreasing, ones (3p13).

Now, one might be tempted to think that this is only a minor interpretative point; but already the deductive progeny of the twin propositions 3p12 and 3p13 – one of the most impressive in the whole of the *Ethics* – indicates that we should not downplay their significance. First, Spinoza's definitions of love and hate (3p13s), both key ingredients in his psychology, are based on 3p12 and 3p13. Second, the propositions are mentioned numerous times, usually to prove propositions concerning the laws governing our emotional life such as '[i]f we imagine that someone enjoys some thing that only one can possess, we shall strive to bring it about that he does not possess it' (3p32). There is an especially important proposition which could not get off the ground without 3p12 and 3p13, namely 3p28:

> We strive to further the occurrence of whatever we imagine will lead to Joy, and to avert or destroy what we imagine is contrary to it, *or* will lead to Sadness. (3p28)
>
> *We strive to imagine, as far as we can, what we imagine will lead to Joy (by p12)*, i.e. (by 2p17), we strive, as far as we can, to regard it as present, *or* as actually existing. But the Mind's striving, *or* power of thinking, is equal to and at one in nature with the Body's striving, *or* power of acting (as clearly follows from 2p7c and p11c). Therefore, we strive absolutely, *or* (what, by p9s, is the same) want and intend that it should exist. This was the first point.
>
> Next, if we imagine that what we believe to be the cause of Sadness, i.e. (by p13s), what we hate, is destroyed, we shall rejoice (by p20), and so (by the first part of this [NS: proposition]) we shall strive to destroy it, *or* (by p13) to avert it from ourselves, so that we shall not regard it as present. This was the second point. Therefore, [we strive to further the occurrence of] whatever we imagine will lead to Joy, etc., q.e.d. (3p28d, the first emphasis added)

At this point, anybody who has trouble with providing a satisfactory account of 3p12 and 3p13 should be alarmed: 3p28 is an absolutely focal proposition of the Spinozistic theory of human motivation and action. In short, 3p12 and 3p13 can be counted among the basic building blocks of Spinoza's psychology and ethics, and hence it is important for any interpretation of the *conatus* principle to show how they are derived from 3p6. Spinoza needs, for his ethics, a principle of innate perfection-enhancement; I think my reading shows that he is entitled to one.[79]

[79] Thus, if by 'perfectionism' one means a moral theory according to which a good life consists of developing or realizing what is central to the human nature to a high degree – this is the 'stripped down' version of perfectionism that Thomas Hurka (1993, see esp. pp. 3–23) defends – then Spinoza no doubt can be counted among the perfectionists in ethics, just as Hurka (1993, pp. 3–4, 23, 25) in fact claims.

But then, one might object, what should we say about the fact that the way in which 3p6 is formulated so clearly resembles the conservation laws of its time? The first thing to note here is that in certain cases or contexts my reading is congruent with the idea of metaphysical inertia. Moreover, this together with the fact that no finality (at least in the Aristotelian sense) is involved, guarantees that it is not in contradiction with Spinoza's mechanistic tendencies. Now, I take it to be a feature characteristic of Spinoza's essentialism that it is sensitive to the enormous differences in complexity of finite entities. We can first consider a very simple thing, for instance a 'simplest body' that Spinoza talks about in the physical digression: it is distinguished from other bodies 'only by motion and rest, speed and slowness' (2le3a2). The existence of such a simple thing seems to be an exceedingly uncomplicated affair; it is difficult to see any other property to follow from its essence than having a certain kind of motion by which it is differentiated from other things. As Spinoza puts it in his correspondence, the contention that 'from the definition of any thing, considered in itself, we can deduce only one property [. . .] may hold good in the case of the most simple things' (*Ep*83; S, p. 958). As a consequence, in their case *conatus* as a principle of perfect essence realization means, insofar as they are considered under the attribute of extension, merely striving to persevere in a certain characteristic motion – which is exactly the same as what would happen according to the inertial interpretation. In such simple cases, there just is nothing else to do, no more perfection or being to be brought about.[80]

Complex entities, for instance human beings, are another matter: a number of properties, presumably corresponding to their level of complexity, can be deduced from their definitions. In the above-cited letter, Spinoza makes it clear that from the definitions of 'real things' (with the obvious exception of the simplest things) not only one but many properties can be inferred. This is illustrated as follows:

> Simply from the fact that I define God as an Entity to whose essence existence belongs, I infer several properties of him, such as that he necessarily exists, that he is one alone, immutable, infinite, etc. I could adduce several examples of this kind, which I omit for the present. (*Ep*83; S, p. 958)

One certainly regrets this omission; it would be highly interesting to know what Spinoza would say about the properties that can be deduced from the definition of a human being.[81] However, the general idea is clear enough,

[80] Cf. Matheron 1988 (1969), p. 28; Schmidt 2009, p. 304.

[81] As he claims that '[t]he human Mind has an adequate knowledge of God's eternal and infinite essence' (2p47), for instance of the attribute of extension, and as adequate ideas are such that they follow from

and when connected to the doctrine of *conatus* as the principle of perfect essence realization it follows that complex things have many things they resiliently strive to bring forth, if they are prevented from acting freely, or from being purely self-determined. Of course, if we consider any such thing in causal isolation, as it is purely in itself, we arrive once again at a strictly inertial scene: the thing in question just remains, in virtue of its own power, in its perfect being, freely as it is, and continues such existence for ever, sempiternally.[82] Of course, this scenario is impossible, for finite things are always subject to passions (4p3, 4p4c). But these considerations show, I believe, that there are enough similarities between *conatus* as the principle of essence realization and the principle of inertia to account for the fact that 3p6 so obviously echoes certain seventeenth-century conservation laws of motion. Moreover, it should be noted that the *exercise* of the striving power itself fits well in the inertial picture because it has no internal limits: a thing 'will always continue to exist by the same power by which it now exists, unless it is destroyed by an external cause' (3p8d). Thus, unlike the Aristotelian powers of natural change which all have an inherent terminus in which their operation ceases,[83] the striving power is such that its exercise goes on without end, if not interfered with.

The present interpretation squares quite well with the troubling 4p72 that claims that '[a] free man always acts honestly, not deceptively' – even when being treacherous would save him from a present danger of death (4p72s).[84] Now, I think that Spinoza's general model of causation and power pushes him strongly towards this position: to the extent that we are free, i.e. unhindered by external causes, we necessarily strive to bring about everything as determined solely by our essence; and since this kind of unhindered mental activity equals reasoning, it is understandable that Spinoza claims us to preserve our being, to the extent that we are free, by doing what reason dictates. To put it slightly differently, if only undeceitful things follow from our essence and being free equals doing what follows from our essence, then, to the extent we are free, we cannot deceive. I take it that by 'an idea of man' we form 'as a model of human nature which we may look to' (4pr; C, p. 545; G II, p. 208) Spinoza means an idea of a human being in

our essences alone, it seems to follow that having clear and distinct ideas of certain attributes are properties that can be deduced from our definition. For an account of how to conceive actions resulting from bodily autonomy, see Huenemann 2008, p. 102.

[82] See 4p4d. [83] For a compact account of this, see Des Chene 1996, p. 21.

[84] For discussions on 4p72, see Bennett 1984, pp. 317–18; Garrett 1990, 1996, pp. 290–5; Rice 1998.

its full and perfect existence; or 'a free man' as an autonomously self-determining rational creature which he discusses in the end of the fourth part.[85]

I believe Spinoza to reason as follows. Power, God's as well as ours, is power to exist and to act. Hence, we do strive to maintain our essence instantiated in actuality, but our being includes more than just that, and so, by 3p6, we strive to improve our level of perfection and insofar as we succeed in this, i.e. insofar as we are perfect and autonomous, we necessarily bring about things derivable from our definition, even if in some extreme cases this exercise of our power of acting would be counterproductive with regard to prolonging the duration of our existence.[86] This is just the way we are built; given that truthfulness necessarily follows from our nature alone, to the extent that we are free, we have precisely as little choice over the fact that we do not lie as a triangle has over the fact that the sum of its internal angles equals the sum of two right angles. This cast of mind finds a clear expression when Spinoza notes at one point, as if in passing, that the striving to preserve itself is 'nothing but the essence of the thing itself (by 3p7), which, insofar as it exists as it does, is conceived to have a force for persevering in existing (by 3p6) *and for doing those things that necessarily follow from its given nature* (see the Definition of Appetite in 3p9s)' (4p26d, emphasis added). We have seen that Spinoza claims, on grounds of the *conatus* thesis, that we strive to be as active as possible – and as already the definition of activity (3d2) requires, all the things we do when we are active are such that they are understandable through and follow from our own nature alone.

HUMAN AGENCY RECONCEIVED

On the present interpretation, finite things are portrayed as centres of causal activity whose 'power to *exist* and to *act*' has *conatus* character in temporality so that each thing battles to stay in existence and to cause effects as determined by its own nature alone. This central facet of Spinoza's geometry of power requires us to reconsider the form explanations of human action must take. In what follows, I also attempt to consider anew the relationship they have to teleology. Recall my way of articulating 3p6: *each and every singular thing is a powerful entity which, when it encounters opposition, strives to exist and to bring about things derivable from its own definition alone.* The task is now to

[85] I would like to thank an anonymous reader for Cambridge University Press for pointing out the relevance of Spinoza's theory of the 'free man' for my interpretation.

[86] Spinoza can allow this presumably because exercising our power of acting equals understanding; and this, in turn, as Garrett (1996, p. 291) remarks, 'makes a greater part of the mind eternal'.

find out how thus understood *conatus* is connected with appetite and related concepts that Spinoza uses to explain human action.

Already a quick look at the way in which Spinoza defines and talks about appetite and desire reveals their intimate linkage to essences. As already noted, striving to persevere in being is not just some run-of-the-mill feature among many others pertaining to finite things; it is the *actual essence* of things themselves (3p7), i.e. the core constitutive of things considered *sub specie durationis*. Indeed, this is the reason why Spinozistic things can be called strivers. Appetite, in turn, is just one term with which Spinoza refers to human striving, namely 'when it is related to the Mind and Body together' (3p9s); desire is conscious appetite, but apart from that layer of consciousness 'there is no difference' between them (3p9s).[87] It is thus completely consistent for Spinoza to hold appetite to be 'nothing but the very essence of man' (3p9s). These two notions connect the *conatus* principle to specific human actions; in Spinoza's idiom, appetite or desire is our essence *insofar as it is conceived to be determined to do something*.[88] Along the same lines, he claims the decisions we make to be really our appetites (3p2s). Now as we have seen when examining the inertial and teleological readings, how one interprets these claims is absolutely crucial for one's understanding of Spinoza's theory of human agency.

To my mind, these passages are most easily interpreted within the kind of framework I have been gradually elaborating, showing the underpinnings of Spinoza's theory of human action to be as follows. It is in virtue of its definable essence that a thing is set up to produce certain effects, and in temporality this essential causal power strives to drive itself through hindrances to bring about only such effects as are derivable from the definition of the thing in question. Now, concrete situations in which human beings find themselves are, of course, of innumerable different kinds, each situation having a specific set of external causes affecting us. By saying 'as each [man] is affected by external causes with this or that species of Joy, Sadness, Love, Hate, etc., [. . .] so his Desires vary' because desire is our essence as determined 'by whatever constitution' we have 'to do something' (3p56d), Spinoza obviously acknowledges that the conditions in which our striving occurs change virtually ceaselessly – and claims that the causal operations to which our essence is determined vary accordingly, as the prevailing circumstances change. And as the quote above shows, Spinoza likes to use the notion of desire when referring to our *conatus* as determined to act in a particular manner.[89] I take it that, strictly speaking, in any actual situation we cannot find *conatus* without its determinations, but

[87] See also defaff1. [88] For this expression, see 3p56d, defaff1. [89] See also Bennett 2001, p. 216.

only essence as constituted in a certain specific manner through particular desires – desires that, of course, necessarily have the *conatus* character.[90]

The aforesaid does not, however, tell us anything about how our striving becomes determined in differing situations. We can begin discussing this by taking heed of a point made by George Molnar: to be exercised or executed, any power requires a *direction* towards something.[91] That an exercise of power can have a direction without having a *telos* seems evident to me. If, for instance, two moving asteroids collide in space, they both surely have physical force which becomes exercised in the collision into a certain direction, but no teleology is involved. And indeed, there seems to be no reason why the point Spinoza makes in the passage above could not be put in terms of directedness: as circumstances change, the *conatus* power is ceaselessly reoriented or redirected. Often this means being directed to those external objects that are critical for keeping the individual's essence instantiated in temporality. To illustrate, sometimes we want to avoid contact with an oncoming projectile, other times we desire to find certain kind of nourishment; each time our striving to persevere in being is, obviously, differently directed. Garrett puts this point very nicely: Spinoza 'construes all *desire* as the direction of this striving onto particular objects';[92] 'particular human desires are simply particular conscious aspects – that is, conscious directions onto more specific objects – of this general appetite for self-preservation'.[93] So, for Spinoza, the basic causal factor underlying all human action is our striving essence, our particular actions or choices being, in the final analysis, determinations of the essence whereby our striving is directed towards something specific.

All this is still rather abstract and does not say much about why or how certain actions instead of some others are 'chosen' or conducted. Knowing how to give an answer to this is, of course, of paramount importance for action explanations. As noted above, I understand the teleological reading of the *conatus* doctrine to offer us roughly the following useful formulation: *If A is likely to promote self-preservation better than any other action within x's power, x endeavours to do A.* How, on my account, does the striving become directed? What sort of principles does essence-originating human behaviour obey?

To give an answer to this, we can begin by noting the dynamism inherent in Spinoza's theory of human action: our striving is ultimately an expression

90 For more on the notion of constitution, see the next chapter. As Pascal Sévérac (1998, pp. 42–3) rightly observes, desire never exists as an abstract 'faculty of desire', but only as singular desires in particular conditions.
91 Molnar 2003, pp. 60, 63. 92 Garrett 2002, p. 127.
93 Garrett 1999, pp. 322–3; cf. also 1996, p. 302.

of God's power, and consequently so are our appetites, volitions, and desires as well. Also, as I will later on further explicate, our emotions influence our actions, and their definition (3d3) involves the concept of power of acting. In consequence, it is understandable that Spinoza's theory of human action can be conceived in terms of powers, their determinations and directions, and resistances encountered. Consider the following body of propositions, most of which we are already familiar with:

> The Mind, as far as it can, strives to imagine those things that increase or aid the Body's power of acting. (3p12)
>
> When the Mind imagines those things that diminish or restrain the Body's power of acting, it strives, as far as it can, to recollect things that exclude their existence. (3p13)
>
> We strive to further the occurrence of whatever we imagine will lead to Joy, and to avert or destroy what we imagine is contrary to it, *or* will lead to Sadness. (3p28)
>
> Sadness diminishes or restrains a man's power of acting (by p11s), i.e. (by p7), diminishes or restrains the striving by which a man strives to persevere in his being; so it is contrary to this striving (by p5), and all a man affected by Sadness strives for is to remove Sadness. [...] [S]ince Joy (by the same p11s) increases or aids man's power of acting, it is easily demonstrated in the same way that the man affected with Joy desires nothing but to preserve it, and does so with the greater Desire, as the Joy is greater. (3p37d)

Now, all of the items above are dynamic in character, and I would argue the following view to underlie them. Our essential power or striving is always fully exercised, and *it pushes towards the direction where it encounters least resistance and most aid* – and then opposes whatever resistance there is and adheres to whatever aid it finds to increase the degree in which our power is used actively, freely, or autonomously. Certain analogies may help us to understand the manner in which this *direction process* takes place (it should be noted that I do not attempt to illustrate the whole nature of our striving with them). One model can be derived from electrodynamics: just as electrical current takes the path of least resistance, the striving of any finite thing is directed to where it is least resisted, or even promoted. Alternatively, we may consider an example taken from optics: when passing from one kind of medium to another, the refraction of a light wave can be calculated by assuming that light takes the fastest route.[94]

Third, and presumably the best, analogy is what may be called the *dam model* of causal power.[95] Think about a dammed river. When a dam

[94] For Leibniz's view on this, see AG, p. 55.

[95] I am grateful to Koistinen and Pietarinen for discussions that have helped me in developing the analogy.

prevents, either wholly or partially, a river from flowing as its own nature determines, the river no longer acts in Spinoza's strict sense: it does not flow freely according to its own nature but is forced to stand more or less still, depending on the preventing power of the dam; and to the extent it is able to flow, it may be diverted from its natural course, depending on the way the dam is constructed. The power by which the river flows is diminished by the dam, or perhaps prevented altogether: a dammed river's power of acting, Spinoza would say, is decreased when the river is under negative passions. Now – and this is the point I want to emphasize here – the flowing of the river is not end-governed, but this does not mean that the water masses' power is not geared towards something definite: water flows to where it is least obstructed, and pushes against any encountered obstacles. Comparing Spinoza's conception of human agency to a dammed river or electricity taking its path may, at first blush, seem strange; but what principled reason could there be in Spinoza's naturalism for not using models derived from natural phenomena to illustrate the way we act? Moreover, the just quoted passages clearly say that human beings are, in essence, *joy-desiring and sorrow-avoiding strivers* – given this, it appears unavoidable that the idea of a causally potent entity taking the path of least resistance captures rather accurately some important aspects of Spinoza's mindset.

As I have already brought forward, a notable line of division runs between *thoughtful* and *unthoughtful* teleology, which is understandable: that we have conscious thoughts surely seems crucial for our behaviour. For instance Curley draws attention to what he regards as the (problematic) move from the simply teleological doctrine of 3p6, which is of the form *If A would help x, x will do it*, to the teleological-*cognitive* doctrine of 3p28 that has the form *If x* thinks *A would help him, x will do it.*[96] Here we can disregard the fact that these propositions are couched in teleological terms and ask, how does 3p6 square with the cases involving conscious thought? What kind of account of this can my interpretation provide; are there in fact two different doctrines, one involving thoughts, the other not? Now recall that Spinoza writes that there is 'no difference' between our appetite (i.e. non-conscious *conatus*) and desire (i.e. conscious *conatus*) (3p9s), '[f]or whether a man is conscious of his appetite or not, the appetite still remains one and the same' (defaff1). Regardless of what, exactly, this means, the message is at least that whatever difference our thoughts make to the way in which we behave – and Spinoza's way of expressing himself in certain propositions implies that they do have some

[96] Curley 1988, p. 164; cf. n. 35 above.

special effect on our actions – that difference cannot cut so deep it could not be accounted for by the doctrine presented in 3p6 and 3p7.

Thus, I would argue that adding thoughts into the picture I have been outlining cannot change the basic idea that each thing strives against any and all opposition to freely realize its own essence. But the way in which this striving is directed or exerted in concrete situations seems to be different in the case of thoughtful and unthoughtful agents. About the latter we can say that they simply take the path of least resistance – provided that they have *any* control over what kind of external causes they encounter – and then resist whatever opposition there is to be resisted. I take it that some unthoughtful things, such as rocks, cannot have any effect on what kind of opposition they encounter; their *conatus* amounts only to resisting the opposition they happen to come up against. Other unthoughtful things, such as plants, have some resources to direct their powers. I find helpful the following passage by Garrett:

> Some individual bodies, however, have far more resources than a rock or toaster for maintaining the distinctive patterns of communication of motion that constitute their continued existence. Specifically, some individual bodies have systems that *register* small differences in their environments and *utilize* the registration of those differences in pursuing bodies and circumstances that will be beneficial to their own preservation while avoiding bodies and circumstances that will be detrimental to it.[97]

The case of thoughtful human agents appears, however, particularly complex: the channelling of our power often takes place through or via thoughts we have of external things. But how exactly does this happen? Here Spinoza's theory takes an interesting turn; according to it, it is our *emotions* (i.e. affects) that indicate the way in which things affect our power:

> By affect I understand affections of the Body by which the Body's power of acting is increased or diminished, aided or restrained, and at the same time, the ideas of these affections. (3d3)

So emotions are ideas that represent the changes in our power of acting: they tell us whether something helps or hinders our striving, whether or not something allows us to bring about effects that follow from our nature alone. As such they are, in effect, *resistance indicators*.[98] We feel joy (or pleasure) when something aids our striving; and we are saddened (or feel pain) when it is resisted. Moreover, our (imaginative) ideas of how things

[97] Garrett 2008, p. 13.

[98] Cf. Garrett 1996, p. 298. Hobbes (*The Elements of Law* 1.7; *EW* IV, pp. 31–4) and Descartes (*The Passions of the Soul* §§ 47–8, 52, 57; CSM I, pp. 345–7, 349–50) can be interpreted as assigning a similar role to emotions.

affect us make emotions arise in us: '[S]o long as the Mind imagines those things that increase or aid our body's power of acting [...] the Mind's power of thinking is increased' (3p12d). 'So long as the Mind imagines anything of this kind [i.e. anything that diminishes the body's power of acting], the power both of Mind and of Body is diminished or restrained' (3p13d). From all this it follows that the directedness of human agents is mediated by emotions,[99] and when we have thoughts or images about external objects, some of those thoughts stir emotions in us, thereby thrusting us either to attain or avoid those objects. This means that the representational content of our ideas of future states of affairs is causally efficacious because content has an effect on how our power is channelled to the direction which *feels* least sorrowful and most joyful.[100]

Of course, our images of external things are inadequate ideas and as such they can, and often do, lead us astray: an apple might be poisoned without our knowing it, and so an (erroneous) idea of the apple's joy-inducing qualities determines our essence to act in a way that, in fact, decreases our power of acting.[101] However, this does not go against the fact that insofar as we are 'in ourselves', we neither act destructively nor are indifferent with regard to our prevailing state, but strive to maximize our power; the basic theory does not become distorted.[102] On the present interpretation, such claims as '[w]e strive to further the occurrence of whatever we imagine will lead to Joy' (3p28) are exactly appropriate.

The aforesaid allows us to draw a more comprehensive picture of human action. Recall, for the last time, the celebrated passage,

> when we say that habitation was the final cause of this or that house, surely we understand nothing but that a man, because he imagined the conveniences of domestic life, had an appetite to build a house. So habitation, insofar as it is considered as a final cause, is nothing more than this singular appetite. (4pr; C, p. 544; G II, p. 207)

From the sort of dynamistic perspective depicted above this is quite naturally read as saying that the essential striving is manifested as an appetite to build a house because accommodation appears as something that gives us

[99] Thus I think that Parkinson (1981, p. 14) is fully correct in claiming, 'pleasure and pain do not create, but *direct* a person's *conatus*'.

[100] I think that in this respect my position agrees with that of Garrett (2008, p. 15): '[I]t is only or chiefly through contributing to the determination of the strength and direction of an individual's *conatus* that an idea can exert power *in* that individual.'

[101] In such a situation, the idea of (eating) the apple first makes us joyful, but then, after having eaten a poisonous apple, we experience pain.

[102] Cf. Jarrett 1999, p. 23.

most joy: the appetite for house-building is the resultant of the human power, the ideas of things surrounding us, and the way in which those ideas make us feel. Of course, Spinoza does not even attempt to develop a detailed science of predicting human behaviour; the general outlines put forward here are enough for the purposes of his ethical project. Moreover, they do allow action explanations of the form, *If, of the options available to x, A least resists x, x strives to do or maintain A.* And in the cases in which human behaviour is determined through ideas, I think the following definition of a mind m seeking after the idea I, suggested to me by Koistinen, is right: *If m in addition to I has an idea B which is such that the level of perfection of m with I is greater than m with I and B, and I and B are mutually exclusive, then m with I and without B will remain.* The formulation of my choice would be, *If, of the ideas present to x, I_A is most pleasurable to x, x strives to do or maintain A.*

ESSENTIALISM AND TELEOLOGY

I have been discussing Spinoza's theory of human action in terms of powers, directions, and resistances. We may now return to the issue of teleology and ask, what should be said about certain Spinozistic claims that seem to allow teleological explanations? In what follows I hope to show that finding the right answer to this question leads to a deeper understanding of Spinoza's action theory.

We can begin the reassessment of Spinoza's relation to teleology by taking a look at the *conatus* principle in its basic form. On my interpretation, it can be stated as follows: *If x encounters opposition, x strives against that opposition to exist and to bring about things that follow from x's essence alone.* I think this allows us to say, preliminarily and loosely, about a particular action, *If x is prevented from bringing about something that would follow from its essence alone, say A, x strives to do A.* These formulations do not appear to have anything teleological in them: the striving any being is endowed with is not explicable in terms of its beneficial consequences, it is just the way in which our power is manifested in the temporal reality of existential 'friction'. Thus 3p6, in itself, seems to be non-teleological.

But what happens when the way in which this striving is directed in concrete situations is taken into account? *If, of the options available to x, A least resists x, x strives to do or maintain A* was my suggestion for the general principle that applies to the directedness of all Spinozistic agents (i.e. taking the path of least resistance); *If, of the ideas present to x, I_A is most joyful to x, x strives to do or maintain A* is its thoughtful variant. These principles govern the way in which the striving of things is directed to something specific. Now, if we define teleology as a style of explanation in which events are explained by their (presumptive) effects, it is obviously possible, based on the aforesaid, to construct teleological action explanations. If someone asks, 'Why does x set itself to doing or maintaining A?', we could reply, basing our answer on Spinozistic principles, 'Because A gives x most joy' or 'Because the idea of the doing of A gives x most joy.' Hence, x's behaviour seems to be (teleologically) explicable by its effect, the bringing about or maintaining of the joy-giving A. Of course, something that gives x joy now may not do so later; and our ideas of external objects are inadequate so that even something whose idea gives us joy now may in fact, once attained, end up making us sad;[103] but the explanations in question are of the teleological form all the same.

To bring things into sharper focus, let us suppose that we know the image of A to give joy to a human agent x; that x knows she can bring about A only by doing B; and that x has enough power to do B. Given Spinoza's theory, we can be fairly sure that x will do B. Only if there is something else available to x, say C, by which x knows she can cause something even more joyful to occur, say both A and Z, would x go for C instead of B. And if queried, 'Why did x do C?' we could reply 'Because x could bring about A and Z by doing C' or even 'x did C in order that A and Z should occur' – answers falling squarely within the teleological camp.

However, even after all this, I would insist that we should not regard Spinoza as a committed teleologist. I think he believed it to be possible to carry out a full-scale reductionist programme in which all statements containing teleological terms or elements are translated into statements referring to non-teleological items only. But as I contend that the sort of non-teleological framework put forward by Carriero, for example, is quite problematic, it is my task to explicate how, then, Spinoza's programme is supposed to work. Before embarking on this, I would like to make one observation: I believe Spinoza would be rather surprised to find himself confronted with questions concerning teleology. The reassured anti-teleological tone of the *Ethics* and its geometry-inspired essentialism suggest that he firmly believed his system to be purged of teleology. And it definitely *is*, if by 'teleology' is meant a certain key feature of the ontological framework that dominated the Renaissance period. But then again Spinoza seems to be, due to this historical context, more or less blind to the fact that teleology in some other sense may have crept onto the pages of the *Ethics*. So, if we were to press Spinoza by insisting that

[103] Cf. n. 101 above.

for instance the often mentioned 3p28, '[w]e strive to further the occurrence of whatever we imagine will lead to Joy', obviously allows teleological explanations of the kind just formulated, what could he say?

I believe Spinoza would consider such teleological explanations very incomplete at best. Let us say that x sets itself or strives to do A. Now, if someone claims Spinoza to be committed to teleology because his system allows explaining x's action by the fact that the idea of the doing of A gives x most joy, Spinoza could retort that this sort of explanation can only be a preliminary one and ontologically speaking not very profound; most importantly, it depends for its validity on a certain ontology of essences, power, and striving – namely, on the geometrical dynamism that is non-teleological by its very nature.

The proper and full Spinozistic explanation of x's action would, I suggest, run roughly as follows. If we know (1) x's essence and what follows from it alone, (2) that x strives against any opposition to perfectly realize its essence, (3) that x directs its power to where it is least resisted (or least sorrow is felt), (4) what is x's prevailing state, and (5) in what circumstances x must operate, we are in a position to provide adequate explanations and predictions of x's actions. For instance, if A can be derived from x's definition and x has been prevented from doing A, it follows that the idea of making A happen gives pleasure to x because A's realization would mean an increase in x's perfection or power of acting; and if the doing of A becomes available to x, x sets itself or strives to do A – only if there is something else available to x, say B, whose idea is even more pleasurable than that of A or of anything else available to x, x would go for B instead of A.[104] This kind of explanation is based on the true causal architecture of the world; feeling pleasure and desiring to act accordingly are the way in which it is manifested in surface phenomenology. The fact that x's nature is as it is explains why it is precisely the idea of A that gives us joy in the first place, or increases our power of acting. And were there no basic causal thrust to preserve oneself, i.e. were there no *conatus*, it is hard to see why any mere idea of the goodness of an action, or any emotion, could prompt us to act. Finally, the fact that the basic striving is channelled to the direction in which least resistance is felt

[104] The full explanation of a case in which an agent does something to attain something else is slightly more complex. For instance, if A can be derived from x's definition, x has been prevented from doing A, and x knows it can do A by doing C, it follows that the idea of the doing of C gives pleasure to x; and if the doing of C becomes available to x, x sets itself or strives to do C. Only if there is something else available to x, say D, whose idea is even more pleasurable than that of C or of anything else available to x, would x go for D instead of C.

explains why x is determined to act in a specific way, or why x desires precisely, say, B and not something else.[105]

This kind of explanations – that succeed, to use Spinoza's expression, in explaining things 'through their first causes'[106] – do not depend for their validity on any prior ends but on essences, what follows from them, and the striving of realization that takes the path of least resistance. Now, especially with regard to human action, it is tempting to resort to teleological explanations of the form 'x did A because it was the most useful thing for x to do', but I think that we stay best true to Spinoza's cast of mind by being clear about the fact that really, deep down, the nature of human action which we are prone to talk about in teleological terms is not in some of its fundamental aspects different from electrical phenomena: despite the fact that the direction processes are of very differing complexity, in all cases entities are thrust to action by their striving to cause effects determined by their essences as freely as circumstances allow. *Any power needs a direction; but it does not need a* telos, nor does Spinoza think it would. Like everything,[107] representational content has causal power, and can be a factor in the direction process, but Spinoza does not see this as leading to teleology. When singular things of geometrical dynamism are cast into the turmoil of temporal existence, they strive to direct their essential power towards least resistance, and that is basically all there is to it.

One revealing point, to my mind not enough emphasized in the literature, is that precisely *the essences give criteria to what can be called 'good' in the first place*; no prior or other standards of good are to be found (or, for that matter, needed). Because striving is the essence of things, and desire is that essence as determined to act, the claim that 'we judge something to be good because we strive for it [. . .] and desire it' (3p9s) means that our essence and what follows from it define goodness. Commentators, especially those offering teleological interpretations of

105 It seems to me that although Spinoza's powers may be depicted by vectors, it is not the case that when there are many different kinds of human strivings competing against each other the resulting course of action would be the sum, as in adding vectors together in mathematics, of competing strivings. For instance, if there is, say, a desire D_1 to move northwards and a stronger desire D_2 to move southwards, the result is not D_3 of strength D_2-D_1 to move southwards but simply acting according to D_2. If I have a desire to take a shower and a slightly stronger desire to go to the kitchen and eat a loaf of bread, the resulting action is not, for example, that I go to the kitchen and drink a glass of water because that is what would result from adding the two desires together. In other words, the winner takes it all, i.e. gets to produce its effect without qualifications, and this is just the way it should be according to the contest view presented above. At least with respect to *human* action this line of thought makes sense, and 3p43d seems to lend support for my interpretation. For especially relevant passages in Descartes, see *PP* 2.45 (CSM 1, p. 244) and *The Passions of the Soul* § 47 (CSM 1, p. 346).

106 For this expression, see *TdIE* § 70; 1p8s2, 2p18s, 3pr (C, p. 492; G II, p. 138), 3p59s; *TTP* III (S, p. 417; G III, p. 46).

107 See 1p36.

Spinoza, seem to downplay this point.[108] However, I would emphasize that the fundamental explanation of things' existence and manner of operation is not to be found in the consequences of their actions but in their essences that constantly bring about effects that *follow* from God's nature, with just as little teleology involved as in geometry (recall the discussion of 1app and teleology). Our essences determine what we strive to bring about, or what we do insofar as we are 'in ourselves', that is, perfect and free – and precisely the items that follow from our essence are the good, mistakenly understood as ends. Just as 'there can be little question that Spinoza rejects the idea that the perfection of God's modes *explains* God's action of creating them in any way',[109] there is no independent goodness or perfection in things or actions to explain the fact that they are good for Spinozistic individuals.[110] In other words, our actions fully determine what is good; what is commonly regarded as self-preservatory or virtuous is of no relevance here.[111] In a sense, it is thoroughly understandable to claim that the nature of a thing determines what is good or useful for it; for instance, what is suitable nourishment for a horse is not necessarily so for us. Thus Spinoza would accept both *If x actively brings about A, it will be good for x* and *If A is good, i.e. if A follows from x's essence alone, x strives to do A.*

I have argued that any doctrine of ends threatens to do what Spinoza emphatically rejects, to reverse the order of nature:

> For what is really a cause, it [the doctrine concerning the end] considers as an effect, and conversely [NS: what is an effect it considers as a cause]. What is by nature prior, it makes posterior. And finally, what is supreme and most perfect, it makes imperfect. (1app; C, p. 442; G II, p. 80)

[108] Jarrett treads an interesting middle ground here: although he argues that Spinoza accepts thoughtful teleology, he contends that 'our most basic desire is the source of all judgments of good and evil' (Jarrett 1999, p. 23) and denies (pp. 8, 15) that Spinoza would have found any kind of functional explanation (i.e. explaining a phenomenon by its effects) acceptable.

[109] Garrett 1999, p. 329.

[110] There is a striking sense in which the Cartesian God and Spinozistic finite beings resemble each other. About the former, Garrett (1999, p. 328) writes:

> [D]ivine creative activity is an example of teleology for Descartes, and the divine actions themselves will have teleological explanations insofar as they are aimed at, and selected in order to produce, their objects. [...] Nevertheless, for Descartes the teleological selection process of the divine will is – unlike that of the human will – an entirely *indifferent* one, not based on the perceived goodness of any consequences (Sixth Objections and Replies; CSM II, 291). In his view, God does not choose things because they are good; rather, things are good because God chooses them to be so.

Leibniz (AG, p. 242; see also Garrett 1999, p. 328) contends that Descartes 'does not want his God to act in accordance with some end'. Now the same holds for Spinozistic individuals: there is no finalistic explanation for *what they strive to bring about*. This suggests that Leibniz would not be prepared, to my mind for understandable reasons, to classify Spinoza's action explanations as properly teleological.

[111] Once again, recall 4p72.

Let us now take a closer look at what he has in mind. To start with priority, recall that the intrinsically valuable ends were traditionally regarded as ontologically prior and something around which causality revolves.[112] How does Spinoza understand priority? At one point he claims that God's nature is 'prior both in knowledge and in nature' (2p10cs); this is understandable because God's nature through which everything must be conceived is the cause of everything. This together with the fact that in 1app Spinoza ties priority with acting as a cause strongly suggests that being prior in nature refers simply to being the cause of something. It is easy to see that in the passage under scrutiny the ends posited by teleological doctrines are what actually is posterior and an effect, but falsely considered as prior and a cause. But what, then, does *really* qualify as prior and a cause for Spinoza? He does not spell it out; but he is most likely implying here that essences, or specifically God's essence, is what is prior, the cause (of everything), and most perfect. I suspect that he might take it that by 1app, it is plain that essences are the sources of causality, and hence omits mentioning them when talking about causes; this may be compared to the way in which he has just similarly 'proven' that 'all final causes are nothing but human fictions' simply by reminding us of 1p16 and related propositions. That essences are prior to what follows from them squares well with the idea that the nature of each thing, as desire, determines what is good for that thing; the existence of goodness of things depends on the essences, but the existence of essences does not depend on the existence of goodness of things. The essence is always ontologically prior and the cause, its effects (that which we strive to bring about, often falsely called ends) are ontologically posterior. This is in line with – and obviously stems from – the Descartes-inspired version of Platonic essentialism explicated in the first chapter.

To conclude, I would argue that this section has established, at the very least, that even if one wants to emphasize the fact that certain propositions of the *Ethics* seem to allow teleological explanations, and even if Spinoza's model is not, in the final analysis, as free from all teleology as its author would like to have thought, it is nevertheless uneasily classified as properly teleological, so decisively does it differ from the conceptions of teleology preceding it. Even more important is Spinoza's positive contribution: an original power-based theory of agency. Finally, Spinoza could legitimately say that since his system is based on the metaphysic of essences that provide the all-important criteria for behaviour, it is not objectionably teleological in the sense that it does not make what is posterior prior.

[112] Aristotle discusses priority and posteriority in his *Metaphysics* (1018b9–1019a14; *CWA* II, pp. 1608–9) and *Categories* (14a27–b23; *CWA* I, p. 22).

CHAPTER 6

Geometrical dynamics of individuality

From the opening chapter of this book, I have argued that Spinoza's theory of individuality is based on a particular brand of essentialism in which geometrical objects are taken as the paradigm for all things. When combined with certain dynamistic intuitions, the result is a view according to which essences make things what they are by determining how God's power is modified so that certain individuals exist. Spinozistic essences thus occupy the esteemed ontological territory traditionally possessed by the substantial forms which had fallen into ill repute. The *conatus* doctrine, in turn, is such an important facet of Spinoza's essentialism that only now, after having a proper grasp of that doctrine, are we in a position to draw a fairly comprehensive picture of the way in which its author thinks about the existence and individuation of finite entities. Perhaps most strikingly, Spinoza wants to elaborate a veritable 'geometry of emotions' that would tell us how certain properties or states are necessarily brought about when a creature with our kind of striving essence is modified by external causes that increase or decrease our power of acting. In what follows, my aim is to explicate the way in which this union of the geometrical paradigm and dynamism allows Spinoza to offer an elaborate account of human existence which takes place under two attributes, thought and extension.

POWER AND INDIVIDUATION

What makes things what they are? This is a basic metaphysical question; perhaps even *the* metaphysical question concerning finite existents. Stated more precisely the question is, what makes reality such that a specific individual exists? Moreover, what explains the individual persistence we encounter in the world? That is, why do things so often maintain their identity and being over time? Aristotelian scholastics were very much occupied with these questions, and they posed some thorny problems to Descartes and Hobbes. I believe that Spinoza patiently crafts an elegant theory of metaphysical individuation. At this point of this study, explicating its main idea can be done rather briefly.

For Spinoza, as for any committed essentialist, essences form the bedrock of individual being; so it is on them that a theory of individuation must be grounded. However, more than just essences is involved here. I have already stressed that in the essence/property ontology, the model of which is provided by geometrical objects, there are certain properties that inevitably accompany the essences. Thus becomes formed what can be called an internal structure a certain individual is endowed with, or the thing's own kind of being. When to this is added the idea that real existence is a causally powerful affair, we end up with powers bringing about effects derivable from definitions that express essences. Thus, *when there is power to determine reality as specified by a particular definition, there is an individual.* This is the general level answer Spinoza's geometry of power can offer to the foremost question of metaphysical individuation.

But what about the question concerning persistence of individuality? Here a number of difficult issues cannot be avoided, for we are dealing not only with a temporal phenomenon but one that involves alterations that do not necessarily equal the termination of existence. In other words, a proper theory of individual persistence must be capable of accounting for the fact that things can persist in existence – and thus remain the *same* – while changing in some notable respects – and thus become *different* from what they were before. Here the notion of *conatus* assumes an eminent task: it is in virtue of having power as striving that things persist in actuality. However, this is only the incontestable point of departure; there is a considerably more sophisticated dynamistic theory to be uncovered, and one in which a notion we have already discussed, that of *power of acting*, plays a key role: as Spinoza states, it – along with our striving – can increase and decrease. An especially telling passage reads:

> [W]hen I say that someone passes from a lesser to a greater perfection, and the opposite, I do not understand that he is changed from one essence, *or* form, to another. For example, a horse is destroyed as much if it is changed into a man as if it is changed into an insect. Rather, we conceive that his power of acting, insofar as it is understood through his nature, is increased or diminished. (4pr; C, pp. 545–6; G II, pp. 208–9)

So an individual's power of acting can change without the loss of essence, that is, without the individual becoming destroyed. This, however, does not prevent Spinoza from saying that through these changes we exist by the *same* power we had when we began to exist.[1] How can we understand all this?

[1] See 3p8d, 4pr (C, p. 546; G II, p. 209).

Now, based on the discussion thus far, it is well-grounded to claim that *as long as there is power that strives to drive through opposition to bring about the kind of effects or states inferable from a definition* (i.e. as long as there is a certain kind of *conatus*), *an individual thing persists in actuality.* On my interpretation, that power individuates things is the focal metaphysical tenet in Spinoza's geometry of power, and now we can say more precisely that when there is a striving power of a specific character, even of a weak intensity or very badly under passions, the individual persists; when that power is taken away, the individual is destroyed. Spinoza says in 2p45s that even if each thing is 'determined by another singular thing to exist in a certain way, still the force by which each one perseveres in existing follows from the eternal necessity of God's nature'. Together with the aforesaid this means, I think, that as our power becomes determined by external causes, the extent to which we are in ourselves may (and does) change – change in the strength of our striving and the intensity of our power of acting – but *the basic character of our striving power (and thus the basic nature of our being) cannot change as long as we actually exist.*

The doctrine of striving as our actual essence thus contains two intertwined features with their respective individualizing tasks. First, that we are powerful strivers is a source of ontological flexibility: the strength or intensity of the *conatus* that keeps us going can be enhanced or diminished. Second, that there is an eternal blueprint of being which our power strives to realize fixes the character of that power, making it the same power through all the victories and losses in the struggles it cannot but engage in. Interestingly, *if* – I would in fact say *as* – Spinoza considers essences to be individual (i.e. unique to their possessors), we all are endowed with our peculiar kind of power that makes us and maintains what we are.[2]

The fact that the power through which things persist has a fixed character should not keep us from appreciating the way in which our actual existence is always subject to alteration; a definite dynamics of being applies to all actual existents. Remember the dynamic essentialism outlined in Chapter 3: there is no potentiality in Spinoza's system, all finite power is always fully in use. What varies is the extent to which our power is that of acting – bringing about effects understandable through our own nature alone – and the extent to which it is exerted passively – expended on bringing about effects as affected by other things' power. Now recall the dam model of causal power:

[2] See especially 3p57 (but also 3p55cd) for the claim that each individual's desire, being its very essence, differs from other individuals' desire as much as their essences differ. Hence, the basic character of our striving or desire is determined by the kind of essence we have.

precluding an effect that would follow, were there no hindrances, from a thing's essence or intrinsic power does not equal destroying its cause, for even when dammed the water masses usually retain their intrinsic power and thus their existence, and start flowing as soon as and to the extent that the gates of the dam are opened, being thereby determined to necessarily cause certain effects. That something follows from the nature of x alone means, roughly, that x has causal power in virtue of which x necessarily acts to the extent its power is not 'dammed'. This formulation reveals one more reason why changes in finite things' causal status are most appropriately understood in terms of changes in power: that x is said to have power to E due to its essence alone does not require E always to be realized, only that in suitable circumstances, that is, if not prevented in any way, x will do E.[3] The dispositional *and* necessary character of finite things' causality is neatly captured by Spinoza's model of power which explains why preventing something that necessarily follows from a finite thing insofar as it is in itself does not need to amount to destroying the thing: as long as the thing's *power* remains, the thing in question is not abolished even when it encounters hindrances.[4] This raises the question around which much of the next section will revolve: how, exactly, are cases of passivity to be explicated? But for our present purposes it suffices to restate that *only taking away the power equals taking away the thing.*

I believe that Spinoza develops his dynamic theory of individuality in order to speak intelligibly about ethically relevant temporal changes taking place in human beings; since we are hindered from free action without this usually amounting to our destruction, we can say that our essences can be freely realized to differing degrees – the very same person, with the same essence, can sometimes live under passions (and, for example, behave foolishly), and sometimes, let us say after having studied Spinoza's philosophy, is able to use his or her understanding to form adequate ideas. Thus, the notions of power and striving allow Spinoza to take into account the dispositionality with which temporal beings are endowed, not to be found when they are considered 'under a species of eternity': insofar as things are free, certain properties or effects necessarily follow from their essences; but the extent to which and the way in which things are free alters in the course

[3] As George Molnar (2003, p. 82) puts it, '[p]owers can exist in the absence or in the presence of their manifestations'. This, I take it, would be Spinoza's approach to the traditional problem discussed in Ch. 2 above (n. 16).

[4] Cf.: '[I]f a thing were truly "in itself" there would be no place for talking about what it would do, if such and such were the case; everything that we say is "in its power" would in fact be done' (Parkinson 1981, p. 10).

of time. As indicated above, I believe that Spinoza saw this kind of dispositionality as innocent of teleology. In the essentialist terms of the opening chapter, it is the formal essence that guides or directs one's actual essence as striving; no ends are needed for this. If we knew a completely active finite thing's – for instance a human being's – definition, we could demonstrate everything the human being in question would do, with just the same kind of necessity as in geometry.

GEOMETRY OF THE PASSIONS

As its title signals, Spinoza's masterpiece is a work with a definite ethical aim: building on a sound ontology, it wants to show us the true nature of our emotions, through the mastering of which we can become creatures that use their power to understand. And understanding is, for Spinoza, the only thing on which lasting happiness can be based; it is the only genuine good. Specifying how to make the transition from passivity to activity, and many related issues, would take us too far afield; my task has been to pin down the ontological engine driving Spinoza's project. There is, however, an important problem yet to be settled, or a question that can be answered only through an explication of an aspect of Spinoza's conceptual machinery on which rests nothing less than his whole philosophical psychology. The problem is as follows. The geometrical paradigm may capture quite effortlessly cases in which a finite temporal existent is active, or in itself: then effects are brought about in virtue of the thing's nature much as certain properties necessarily follow from the nature of a geometrical figure. Geometrical properties are precisely *necessary* properties (in traditional parlance, *propria*): taking away any such property makes its possessor vanish. Now real things of the temporal world change constantly, their intrinsic power being 'dammed' in various ways so that effects (i.e. properties) they would have caused had they been free are not realized, and something else is done instead. But how, in light of the geometrical model, can this be the case? Does not preventing something which follows from a nature insofar as it is unhindered inevitably result in destroying the thing in question?[5] And even if it were granted that, obviously, things can become passively determined in many ways without being thereby destroyed, how can we account for passions (such as passive emotions) in a way that would meet the rigour of the geometrical paradigm? Is the geometrical model in

[5] In alternative terms: even if God's causality, or the following from God's nature, is of the same kind as the one we find in geometrical entities, how should we understand the dispositionality characterizing temporal things' but not geometrical entities' being?

fact inapplicable here?[6] I will answer these questions through an analysis of Spinoza's geometrized way of analysing and taxonomizing passive emotions. However, the following discussion is directly relevant for the correct analysis of human action (both active and passive, mental as well as physical), too.

It really is no exaggeration to talk about Spinoza's philosophical psychology as a geometry of emotions. The famed preface to the third part of the *Ethics* leaves little room for doubting that developing such a brand of knowledge – considering 'human actions and appetites just as if it were a Question of lines, planes, and bodies' – is his *goal*; whether one believes this to be a feasible undertaking is another matter. It should be noted that, for Spinoza, this is a deep-rooted conviction, making its presence felt in the *Theological-Political Treatise* in a revealing manner:

> Nobody can so completely transfer to another all his right, and consequently his power, as to cease to be a human being, nor will there ever be a sovereign power that can do all it pleases. It would be vain to command a subject to hate one to whom he is indebted for some service, to love one who has done him harm, to refrain from taking offence at insults, from wanting to be free of fear, or from *numerous similar things that necessarily follow from the laws of human nature*. (*TTP* XVII; S, p. 536; G III, p. 201, emphasis added)

In the *Political Treatise* the point is made more emphatically still:

> Therefore in turning my attention to political theory it was not my purpose to suggest anything that is novel or unheard of, but only to demonstrate by sure and conclusive reasoning such things as are in closest agreement with practice, deducing them from human nature as it really is. And in order to enquire into matters relevant to this branch of knowledge in the same unfettered spirit as is habitually shown in mathematical studies, I have taken great care not to deride, bewail, or execrate human actions, but to understand them. So I have regarded human emotions such as love, hatred, anger, envy, pride, pity, and other agitations of the mind *not as vices of human nature but as properties pertaining to it* in the same way as heat, cold, storm, thunder, and such pertain to the nature of the atmosphere. These things, though troublesome, are inevitable, and have definite causes through which we try to understand their nature. (*TP* I.4; S, p. 681; G III, p. 274, emphasis added)

So, all the emotional states we *undergo* – feeling hurt when insulted, being grateful to those who have helped us, and so on – necessarily follow from the way in which we are built; clearly, in Spinoza's mind, there is no reason to think that our passions could somehow escape the iron-clad geometrical necessity structuring the rest of the world.

Surprisingly little attention has been paid to a notion involved in Spinoza's attempts to unlock these issues, namely that of *constitution*

[6] I would like to thank an anonymous reader for Cambridge University Press for pressing this issue.

(*constitutio*). In fact, passages in which the term figures are so numerous in the *Ethics* that listing them would not make sense. Clearly, the notion is designed to be used in accounting for change in temporal existents. The main idea is that *actual essences are constituted varyingly over time*; the way in which an essence is constituted seems to determine what Spinoza calls the actual being of a thing.[7] I find absolutely central the following passage we have already touched upon:

> But Desire is the very essence, *or* nature, of each [man] insofar as it is conceived to be determined, by whatever constitution he has, to do something (see p9s). Therefore, *as each [man] is affected by external causes with this or that species of Joy, Sadness, Love, Hate, etc. – i.e., as his nature is constituted in one way or the other*, so his Desires vary and the nature of one Desire must differ from the nature of the other as much as the affects from which each arises differ from one another. (3p56d, latter emphasis added)

So, when we are affected by external causes, passive emotions often arise; this is said to equal our nature being constituted in different ways.

We may use the ultimate Spinozistic entity of illustration to build up a very robust example of how Spinoza thinks about these matters. Think of a triangle; it can be defined as *a closed plane figure formed by three intersecting lines* – a definition which expresses an essence (it is a description of essence-constituting 'triangularity'). Further, let us suppose that the triangle is a completely isolated actual existent and one with the property of *having only 60 degree angles* (we would call such a triangle equilateral, with all sides of equal length) (Fig. 2). Now let us place it into the causal network formed by actual finite things so that it remains no longer unaffected by external causes; it becomes determined in various ways. For instance, it may be 'squeezed' so that only two of its sides are equal (in which case it would be called an isosceles triangle) (Fig. 3). Now, it is still a triangle, a closed plane figure of

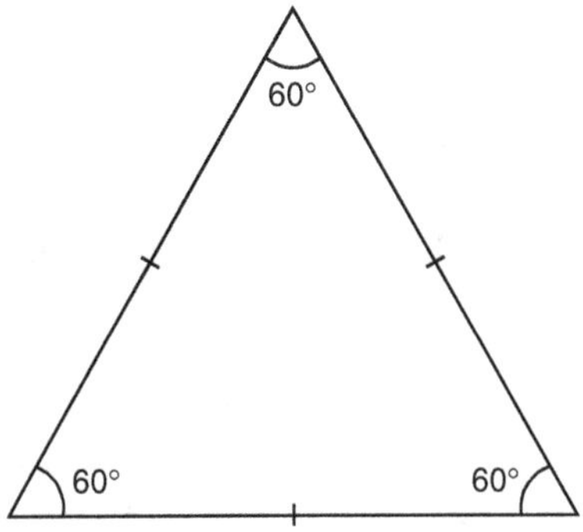

Figure 2

[7] See e.g. 2p11 together with 3p3d.

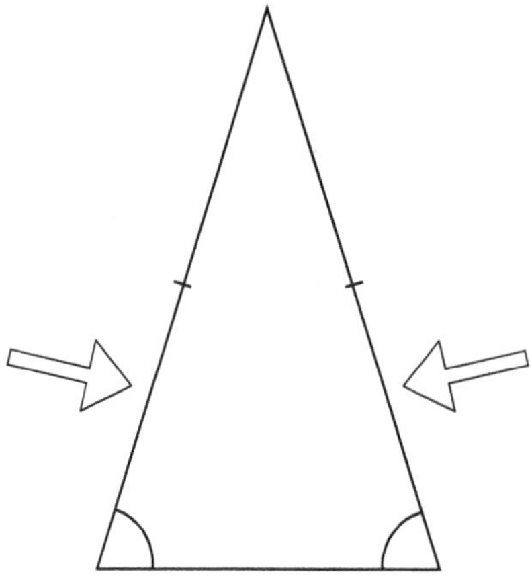

Figure 3

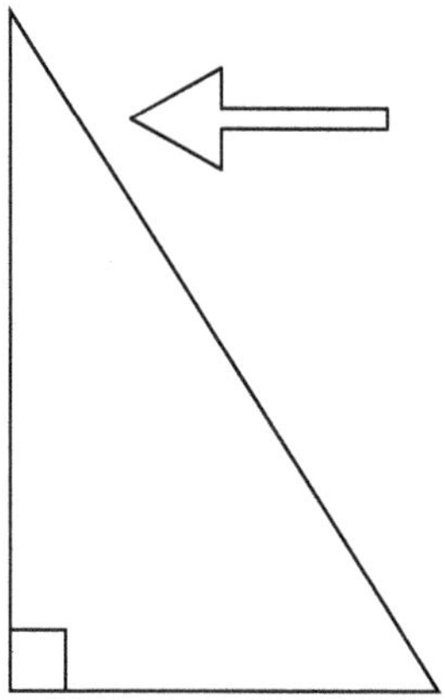

Figure 4

three sides – its essence has not changed into something else. But now the essential principle of 'triangularity' is constituted or 'put together' differently, with accompanying change in properties. The property of having only 60 degree angles is now gone and a new one has emerged, that of *having two equal (base) angles*. We can take the illustration further; let us assume that the triangle in question becomes 'tilted' in a way that one of its angles equals a right angle (Fig. 4). Now many properties disappear and at least the property of *fulfilling the Pythagorean theorem* appears. Again, the essence ('triangularity') remains intact, but it is constituted anew, from which constitution new properties follow.

At this point a few observations are in order. First, when the triangle illustrated is determined by external causes, the various properties *follow with the same kind of necessity* as they do in the case in which the triangle is completely self-determined (when it is in the equilateral constitution). Second, provided that the triangle is an actual Spinozistic existent, there is all through its temporal existence *a striving to be an entity with those properties only that belong to the equilaterally constituted triangle* (e.g. the property of having three 60 degree angles). Third, in each case there is a specific type of causal process (a 'proximate' cause) which generates a certain constitution and a set of properties.[8] I firmly believe that this faithfully captures Spinoza's mindset when he goes on to build his theory of emotions. The illustration has, of course, some peculiarities we should be aware of (it is, after all, an illustration, not the illustrated thing itself); most notably, we all seem to be, to speak figuratively, our own *kinds* of (actual) geometrical entities (thus comparable to triangles, squares, circles, etc.).[9] Moreover, our essences are much more complex than that of, say, a triangle. However, Spinoza thinks there is enough common in human essences, or essential principles, for us to share certain properties (e.g. emotions), which arise from clearly specifiable interactions with external causes.[10] The illustration above is designed to reveal the core of the geometrical architecture underpinning such causal entities as emotions that Spinoza is determined to analyse and classify. Indeed, it is in virtue of this architecture that a science treating emotions as if they were 'lines, planes, and bodies' is a possible project to begin with – *we are just dealing with entities endowed with causal power to realize or 'fill' their clearly definable internal structures.*

The above is, I would argue, what Spinoza has in mind when he talks about emotions arising from us being affected by external causes so that our essence is constituted anew. On the present interpretation, statements such as the following make perfect sense:

> All our strivings, *or* Desires, follow from the necessity of our nature in such a way that they can be understood either through it alone, as through their proximate cause, or insofar as we are a part of nature, which cannot be conceived adequately through itself without other individuals. (4app1; C, p. 588; G II, p. 266)

[8] Koistinen (forthcoming$_a$) defends this kind of conception of the proximate cause.

[9] Cf. Leibniz, who thinks that each substance is 'a lowest species' (AG, p. 42) and thus strictly speaking an entity of its own kind.

[10] As a matter of fact, it is quite easy to see how many emotions – especially the basic ones – can be common to things with individual essences. For instance joy, which is what we experience when we become more perfect (3p11s), is clearly what *any* conscious individual, regardless of the specific character of its essence, feels when it obtains a constitution closer to the self-determined one.

To take a simple, and basic, example of the way in which Spinoza's analysis of emotion works: hate is defined as '*Sadness with the accompanying idea of an external cause*' (3p13s). And because sadness is what we feel when the mind '*passes to a lesser perfection*' (3p11s), this means that when we think that something external to us does us harm, our nature is inevitably so constituted that the feeling of hate follows.

The geometrical model applies even to the emotions *themselves*, for they have essences, too, from which certain things follow.[11] In fact, mapping these structures is something Spinoza considers a central task, as the following remark he makes after having examined issues pertaining to pride – the discussion on emotions nearing its end – reveals:

> *These things follow from this affect as necessarily as it follows from the nature of a triangle that its three angles are equal to two right angles.* I have already said that I call these, and like affects, evil insofar as I attend only to human advantage. But the laws of nature concern the common order of nature, of which man is a part. I wished to remind my readers of this here, in passing, in case anyone thought my purpose was only to tell about men's vices and their absurd deeds, and not *to demonstrate the nature and properties of things*. For as I said in the Preface of Part III, I consider men's affects and properties just like other natural things. And of course human affects, if they do not indicate man's power, at least indicate the power and skill of nature, no less than many other things we wonder at and take pleasure in contemplating. (4p57s, emphases added)

In cases like these, things often become more complex and Spinoza's analyses longer. This holds for pride, which is a joy – what the mind feels when its perfection is increased – that arises when we have an overly high opinion of ourselves (defaff28, 4p57d); because we are also joy-desiring creatures (3p12), someone who is proud strives to maintain or even promote things that keep him feeling proud. For this, judicious companionship will not do, and thus '[t]he proud man loves the presence of parasites, *or* flatterers, but hates the presence of the noble' (4p57).[12] This really must be the high point of Spinoza's rationalistic dynamism: even the most tortuous feeling of ours cannot but yield to the lucidity of an analysis conducted along geometrical lines.

If the present interpretation is accurate, it follows that there are a number of ways in which each thing can be constituted, each way resulting in certain properties and thus in being determined in a certain clearly specifiable and

11 This seems to be in keeping with the tradition, for many scholastics thought that accidents can be individuals (see Carriero 1995, pp. 258–9) and hence, given the Aristotelian metaphysical framework, obviously entities which have their own essences.

12 For a helpful catalogue of the affects, see LeBuffe 2009, pp. 204–5.

necessary fashion.[13] Hence there is in fact *an array of constitution/property structures* for each individual; one of those structures, when the essence is constituted unaffected by external causes (in such a case the individual would be completely self-determined), is the ontologically privileged one corresponding to the essence/property structure of the formal essence (as explicated in Chapter 1). As Spinoza puts it, 'by an affection of the human essence we understand any constitution of that essence, whether it is innate [NS: or has come from outside]' (defaff1).

In terms of the illustration above, our life could perhaps be said to be, for the most part at least, going through sequences of varying *scalene* constitutions, each with its specifiable properties; as we are unaware of many of their properties, or at least of their principled dynamics, Spinoza is confident that his project can greatly enhance our self-understanding. Moreover, all the constitutions are obviously actualized in certain times and places, as determined by God's nature as conceived *sub specie aeternitatis* (or as I would also say, by our formal being); *strivings of actual entities are exercised so that the eternal framework unfolds in temporality.* The theory of constitution offers Spinoza the flexibility required to account for change – the individual persists, but is being constituted in various ways.[14] For each human being, of its innumerable constitutions there is one which is ontologically privileged; it is the perfect way of being one strives to attain as efficiently as one can – and would succeed in attaining, were one's power not always, to a certain extent, hindered. There is thus, I would argue, a striving for an 'autonomous constitution', a state in which everything pertaining to us is self-determined.

Given the aforesaid, not all constitutions are equally desirable in Spinoza's geometrical moral psychology: the autonomous constitution is the *good* one, and what is perceived to lead us closer to it feels joyful. According to Spinoza, that state is one of pure understanding, or the

[13] Here I have been focusing on the attribute of thought, but the doctrine of constitution applies to extended things as well: 'For by an affection of the human essence we understand any constitution of that essence [. . .], whether it is conceived through the attribute of Thought alone, or through the attribute of Extension alone, or is referred to both at once' (defaff1). For bodily constitution, see the next section.

[14] The standard Aristotelian account of non-destructive alterations is that a certain form–matter composite prevails and takes on new accidental forms. Although causally efficacious essences occupy a place in Spinoza's conceptual field that has much in common with the one possessed by substantial forms in scholasticism, Spinoza's essentialism differs so decisively from the Peripatetic one that there is no room in it for anything even remotely resembling accidental forms. The novel theory of constitution allows Spinoza to say that when a thing changes, what happens is that its actual essence is constituted anew, as dictated by God's nature (consider also here how non-finalistic Spinoza's scheme is).

constitution in which our minds bring about adequate ideas only.[15] Moreover, each constitution seems to correspond to a specific degree of activity. Accordingly, the 'psychotherapeutic' techniques for controlling passive emotions, presented in the beginning of the closing part of the *Ethics*, can be seen as strategies of reason through which we can gain control over the way in which our mind is constituted,[16] thus enabling us to increase our freedom, autonomy, and power of acting. This is what we as joy-desiring and sorrow-averting strivers cannot but want.

As we have seen many times over, Spinoza speaks about things that 'follow from our nature'. Somewhat surprisingly, he also does so in connection with passions; recall that 'we are acted on when something happens in us, or *something follows from our nature*, of which we are only a partial cause' (3d2).[17] But how can any passion 'follow from our nature' when it is by definition something for which external causes, too, are causally responsible? I would suggest that the present interpretation offers us a promising solution to this problem. Full understanding of *a specific phenomenon* of causal determination requires knowledge of the particular external causes involved; consider:

> Insofar as a man is determined to act from the fact that he has inadequate ideas, he is acted on (by 3p1), i.e. (by 3d1 and d2), he does something which cannot be perceived through his essence alone, i.e. (by d8), which does not follow from his virtue. But insofar as he is determined to do something from the fact that he understands, he acts (by 3p1), i.e. (by 3d2), does something which is perceived through his essence alone, *or* (by d8) which follows adequately from his virtue. (4p23d)

Prima facie, because effects are conceived through their causes (1a4), this contradicts the definition of passivity in 3d2. However, when the causal interaction has resulted in an essence being constituted in a certain way, certain properties *do* necessarily follow from that constitution, as 3d2 suggests. The essence, with the constitutions it can take, is thus always the causal (and epistemological) centre of gravity.[18] I take it that this is why Spinoza also says that 'even if each one is determined by another singular thing to exist in a certain way, still the force by which each one perseveres in existing follows from the eternal necessity of God's nature' (2p45s). Hence, in dynamic terms, it is always striving power of a certain kind – one directed

[15] For Spinoza's grounds for this claim, see the final section of this chapter. [16] See 5p20s.
[17] Emphasis added.
[18] In terms of the dam model of causation: the results of damming depend as much on the nature of the dam as on the dammed thing's nature.

to a definite constitution of autonomous being, I would argue – that is at work.

The aforesaid has important consequences for Spinoza's theory of action. Given that our actions (in the loose sense that also comprises the passive but conscious things we do) equal us desiring something – desires equalling our *conatus* being directed towards certain objects – and that the desires, in turn, depend on our current constitution, it follows that the above geometrical constitution analysis offers a way to develop what may be dubbed a geometry of human action: we are dealing with specifically constituted loci of causal power. With each constitution, we operate in a particular way, desire some specific object (3p56). Given that striving is our actual essence, when that essence is constituted anew the character of our striving is, understandably enough, altered; in other words, essential causers as we are, a change in our essence's constitution results in change in causation. In light of this, anything but the denial of free will (2p48) would come as a surprise.

Connecting Spinoza's geometrized conception of passivity to his power-based theory of individuation allows us to appreciate his admirably keen philosophical eye. Identifying actual essences with a definite kind of (striving) power allows Spinoza to develop a theory in which the fully autonomous realization of essences can be prevented without this necessarily amounting to the destruction of the things with those essences: as long as there is the striving to bring about properties pertaining to the formal essence, the thing persists, if perhaps only in a very passive state. Even if that power were completely prevented from acting freely, the individual in question can persist, only in a completely passively constituted state; but if that power is taken away, the individual ceases to exist; and I would suggest that to the extent one's striving overpowers external causes, it draws nearer to its active constitution that corresponds to its formal essence.

PHYSICAL INDIVIDUALITY AND THE FIELD METAPHYSIC

It has been convincingly argued by Bennett that Spinoza must allow, despite his property dualism, certain transcategorial concepts that pertain equally to all attributes.[19] Not only logico-numerical and temporal concepts should be counted among the transcategorial ones, but also causal concepts, for 'there must be causal laws operating within each category, and so causal

[19] Bennett 1984, pp. 41–7.

concepts must be protected from the dualist split'.[20] Since Bennett sees the concept of power as a member of this group, it is consistent, and correct, for him to claim it to be applicable under all attributes.[21]

Despite the transcategorial character of causal concepts, it is obvious that power and causation under different attributes mean different things due to the fundamental differences between attributes they are applied to. Spinoza gestures towards this by claiming, 'the Mind's striving, *or* power of thinking, is equal to and at one in nature with the Body's striving, *or* power of acting' (3p28d).[22] We have already encountered 'power of acting' – how does it relate to 'power of thinking' (*potentia cogitandi*)? The following passage provides the crucial clue:

> Suppose the Mind is affected by two affects at once, one of which neither increases nor diminishes its power of acting, while the other either increases it or diminishes it (see p01). From p14 it is clear that when the Mind is afterwards affected with the former affect as by its true cause, which (by hypothesis) through itself neither increases nor diminishes its power of thinking, it will immediately be affected with the latter also, which increases or diminishes its power of thinking, i.e. (by p11s), with Joy, or Sadness. (3p15d)

In other words, the mind's power of thinking is just another name for its power of acting.[23] But these remarks are only preliminary observations on how power under different attributes should be conceived. The discussion of passions above focused on such mental items as emotions; in what follows, I endeavour to present some ideas concerning the way in which Spinoza's views on extended substance and physical existence can be approached using the concept of power as the basis of interpretation. I will first offer a brief outline of Bennett's much-discussed 'field metaphysical' interpretation of Spinoza's doctrine of one extended substance, for I believe that Bennett's approach enables us to deal with the central questions concerning physical individuality in a rather effective way.[24] After having expounded the philosophical problems Bennett's field metaphysical interpretation solves, I will show how his position can be combined with the dynamistic line of interpretation. My main question will be, what follows if physical individuals are seen as parts of a unified field of extended power?

[20] Bennett 1984, p. 44. [21] Ibid., pp. 74, 287. [22] See also 3p11, 3p12d.

[23] Cf.: '[I]t is clear that the mind's power of thinking is its power of acting' (Della Rocca 1996, p. 261).

[24] Bennett 1984 (ch. 4) is here the key text. For other sources that to my mind at least suggest a field metaphysical interpretation of Spinozistic extension, see Wolf 1974 (1927), pp. 21–4; von Dunin Borkowski 1933, pp. 97–101; Harris 1972, p. 195, 1973, pp. 50–6, 65, 68–9; Sachs 1976; Lachterman 1978, p. 103; Van Zandt 1986, pp. 255–60, 264. For a discussion of Spinozism and Bennett's interpretation as space-holism, see Esfeld 2002, pp. 213–26.

Bennett calls substance expressed under the attribute of extension, or the one extended substance, *space*. It must be partless, because if an entity consisted of (really distinct) parts, it would be ontologically dependent on those parts, and hence no longer a substance, given that causal and conceptual independence was a central seventeenth-century requirement for substances. This creates difficulties especially in the case of extended substances, since it was commonly held that every physical object is always divisible into separate parts.[25] Moreover, as Gueroult notes, infinity and divisibility were traditionally thought to exclude each other, and the idea of infinite extension was rejected. But this was not Spinoza's approach; he holds on to infinity and rejects divisibility.[26] What Bennett calls 'field metaphysic' should be seen as a solution to the problems pertaining to the monistic idea of extension as substance: if there really is only a single extended, ontologically independent and unified substance, how can we explicate the relation between physical bodies and space as a whole?

In the important 1p15s, Spinoza makes a distinction between two ways of conceiving extended nature. He claims that those who are inclined to divide extension attend to it only superficially, as it appears to our senses, and consequently find it to be composed of parts. But if the intellect attends to extended nature as a substance, Spinoza insists that 'it will be found to be infinite, unique, and indivisible' (1p15s; C, p. 424; G II, p. 59). Thus, Spinozistic space is a unity: it cannot be divided in the sense of having really distinct parts. According to Bennett, Spinoza can defend this idea of unified space, not constituted by its parts, in the following way:

> [T]he thought of a whole space is not built up out of thoughts of its subregions, whereas the thought of any finite region of space must involve the thought of a larger region – or the whole of space – within which it is embedded. [...] [T]he very concept of a spatial region involves the concept of the space of which it is a region.[27]

In other words, no part of space can exist, or be understood, without relation to the space as a whole, and hence its parts cannot be really distinct from each other. So if unified space is the only substantial extended entity there is, what should we say about finite things? For Spinoza they of course are modes, that is, affections or states of substance that can neither be nor be conceived apart from substance. Consequently, although extended substance cannot strictly speaking be divided, it is not altogether without distinctions:

[25] Bennett 1984, pp. 82–6. See also Bennett 1996, pp. 62–3; Des Chene 1996, p. 99.
[26] Gueroult 1968, pp. 214–15. [27] Bennett 1984, p. 86.

[M]atter is everywhere the same, and [. . .] parts are distinguished in it only insofar as we conceive matter to be affected in different ways, so that its parts are distinguished only modally, but not really. (1p15s; C, p. 424; G II, p. 59)

In other words, and in keeping with what we have seen above,[28] material phenomena result from the one substance being affected in infinitely many ways. On the field metaphysical interpretation, this means that the regions of the whole space can become many different kinds of qualities in such a way that statements about material bodies can be reduced to statements about space. More exactly, any particular body must be a logical construction out of a string of place-times, for Spinoza's basic ontology does not contain physical objects. According to Bennett, this makes bodies adjectival to regions of space; all the statements about bodies can at least in principle be expressed in terms of 'how Space is'.[29] As a result it follows that what we see as a distinct material thing, for example as a chair, is more adequately conceived as a certain spatial region being 'chairy*', that is, having that property of space which we *conceptualize* at the physical level – 'one level up' from the field metaphysical one – as a chair.[30]

According to this kind of field metaphysic, the destruction of a finite individual does not amount to the annihilation of a certain region of extended substance, but only to spatial alteration. This is in complete agreement with the following important passage of the *Ethics*:

For example, we conceive that water is divided and its parts separated from one another – insofar as it is water, but not insofar as it is corporeal substance. For insofar as it is substance, it is neither separated nor divided. Again, water, insofar as it is water, is generated and corrupted, but insofar as it is substance, it is neither generated nor corrupted. (1p15s; C, p. 424; G II, pp. 59–60)[31]

In other words, despite the fact that material things can be driven out of existence, space itself cannot be annihilated; or as Bennett puts it, 'nothing goes out of existence, but something alters'.[32]

Bennett's field metaphysical interpretation has received its share of criticism, but I think it handles very well the texts presented above which are some of the trickiest passages Spinoza ever wrote. Moreover, there is a genuine philosophical problem – namely, how should the character of finite physical entities be understood in substance monism that applies to extension? – that can be answered by interpreting Spinoza the way Bennett does.[33]

[28] See esp. Ch. 3. [29] Bennett 1984, pp. 88–91, 95–6, 2001, pp. 143–4.

[30] Bennett 1984, pp. 96, 106–7, 1996, pp. 70–1. [31] See also *Ep*4; S, p. 767; G IV, p. 14.

[32] Bennett 1996, p. 71.

[33] Moreover, as Bennett (1984, pp. 97–103, 104–6, 2001, pp. 144–5) points out, this line of interpretation also provides solutions to such troublesome philosophical problems as corporeal annihilation and co-location of bodies.

Indeed, it seems obvious to me that Spinoza's thinking about these matters remained relatively stable, and in a way that invites the field metaphysical interpretation, at least since the early *Short Treatise*.[34]

Bennett notes that field metaphysic does not necessarily lead to physics of material particles, and he suggests that the best way to understand what is meant by an 'occupant of space' might be to replace the concept of a physical thing with the concept of a force or a wave.[35] True as this is, I think the question can also be posed the other way round: instead of asking what kind of *physics* can be combined with field metaphysic, it can be asked what kind of view of *the basic nature of substance* can be combined with the field metaphysical character of extension. Consequently, I would like to examine how well the dynamistic line of interpretation fares in connection with field metaphysic.

I have argued, in Chapter 3, that Spinoza's God has power to produce himself and all finite existents in virtue of his essence alone. In this kind of dynamic essentialism, God's nature-constituting attributes must be intrinsically causally efficacious, each attribute capable of bringing about all the things that fall under it (recall especially *Ep*81 and *Ep*83). This, of course, applies to the attribute of extension as well – and fits very nicely together with the field metaphysical interpretation. When combined, the result is the idea of space as *one unified field of spatial power*. Thus, seen from the dynamic point of view, extended power forms a spatial continuum with no really distinct parts. God's essential power is a fundamental unity; consequently, in ontological terms it is not correct to refer to finite bodies without relating them to the whole of God's power, without seeing them as parts of a unified field of causal activity.[36] I would be willing to accept the core of Bennett's thesis: regions of space can be modified in many ways and from this result the innumerable phenomena of the physical world. For my approach, however, the pressing question concerning extended substance and its modes is, what kind of states or affections could there be in *spatial* power?

Now substance as power is no static thing, and its affections or expressions are best seen not in the mould of the properties of everyday things, but as specifically determined centres of causal power, indistinguishable from a certain way of causing effects.[37] Since these modally distinct parts constitute differing amounts of God's power, I suggest that under the attribute of extension we are dealing with differences in power distribution in the spatial

[34] See *KV* I.2; C, pp. 70–2; G I, pp. 24–6. [35] Bennett 1984, p. 106. Cf. Barbone 2002, p. 97.
[36] Cf. 4p4d. [37] See Ch. 3.

field: finite entities are, field metaphysically speaking, constituted by differences in the intensity or strength of spatial power. These *distributive differences in intensity* form relatively stable *spatial patterns* or *formations*, i.e. modes of extension. This is what it means for them to be parts of substantial power. There is no region without power, since that would equal total annihilation of extension, but there are drastic differences in the intensity of power between spatial regions, and out of these differences finite things are constituted.

Hence, the material bodies we encounter in the physical world can be described as modifications of spatial power; from this kind of metaphysical perspective, there are only spatially distributed alterations in power. Consequently, motion and change of a physical body are alterations in the one extended field of power – alterations in which regions of power form such a pattern of intensity that a certain material body is thereby constituted. And following Bennett's analysis, a finite body's destruction cannot mean annihilation of power, but only such an alteration in it that no region carries anymore the formation of intensity that equals the mode ontologically described. Hence, from this viewpoint claims about material bodies are, deep down, not just claims about space, but claims about spatially distributed intensities of power. It is thus preferable to see the field as a field of intensification patterns, not of some other kind of properties;[38] this, in turn, provides us with a view of extension as a spatial field in which modifications as strivers resist and help each other.

The ideas presented here are not as uncommon as one may at first sight be tempted to think. On the one hand, the idea that extended nature is intrinsically active or powerful has a notable seventeenth-century proponent in Gassendi who, as Antonia LoLordo explains, defended the claim that atoms are endowed with *vis motrix*, or motive power, on which the causal efficacy of the material world is based.[39] On the other hand, different kinds of dynamistic field metaphysic are espoused – apart from Kant, to whom Bennett, too, refers[40] – by at least such later thinkers as Schopenhauer – who provides, I think, a field theoretical theory of the material world[41] – and Nietzsche, who in his unpublished manuscripts sketches a picture of reality as a totality constituted by power quanta in a

[38] See also Garrett 2002, p. 156. [39] LoLordo 2005, pp. 75–81.

[40] Bennett 1984, p. 91. According to Jeffrey Edwards (2000, p. 1), the 'thoroughgoing continuity in the reciprocal action of all perceivable substances' that Kant discusses in the Third Analogy of the *Critique of Pure Reason* 'is possible only if space presents us with an all-inclusive or universal dynamical plenum'.

[41] I argue in detail for this in Viljanen 2009a.

relation of tension to each other.[42] Moreover, it is by no means out of place or strange to interpret Spinoza as developing one kind of field theory: if we start from the idea of some basic stuff – be it, for instance, mass, temperature, electrical strength, or power – that is distributed continuously and variably throughout the unique and unified extension, we arrive at a conception of spatial reality as a field.[43] Spinoza can be interpreted as making original metaphysical and ethical use of a notable modern line of thought.

An essence or nature becomes realized or expressed differently under different attributes. In the previous section, we discussed mental passions, a central feature of human life as it takes place under the attribute of thought. But how exactly should we understand corporeal entities, given the field metaphysical interpretation? The 'Physical Digression' of the second part of the *Ethics* (inserted between 2p13 and p14) is the unquestionable basis of Spinozistic physics of matter in motion. Spinoza presents in it interesting claims about a class of physical beings he calls 'individuals', and thus it tells us much about what he takes physical things to be. At this point it should be emphasized that Bennett's talk of Spinoza's two levels, the metaphysical and the physical, concerns two different ways of *describing* the same thing, namely the attribute of extension – although it should be remembered that, in philosophical terms, the field metaphysical description is the far more adequate one.[44]

The famous definition of 'individual' reads as follows:

When a number of bodies, whether of the same or of different size, are so constrained by other bodies that they lie upon one another, or if they so move, whether with the same degree or different degrees of speed, *that they communicate their motions to each other in a certain fixed manner*, we shall say that those bodies are united with one another and that they all together compose one body *or* Individual,

[42] See e.g. *WP* § 635. More recently, Harré and Madden (1973, p. 223) have defended the idea of dynamic fields that comes remarkably close to the present interpretation of Spinoza's thought: '[A] field of potentials is the most plausible candidate for an entity whose powers are its nature, so that if there are any entities which are fundamental then the field is the one most likely to be so[.] [. . .] Fields are structures of spatially distributed potentials.' These potentials, in their turn, 'are the causal powers of points' (p. 224) which are possessed because the field has a certain nature at those points. Harré and Madden even entertain a theory of one universal and infinite field that has internal differentiations within it and hence a structure 'derived from the differentiation of potentials in space and through time' (p. 226). Finally, according to Jammer (1999 [1957], p. 69) 'the notion of a *field* of forces' is anticipated already by John Buridan.

[43] Here I am leaning on Mark Wilson's (1998, p. 668) formulation: 'A physical quantity (such as mass, temperature or electrical strength) appears as a *field* if it is distributed continuously and variably throughout a region.'

[44] Bennett 2001, pp. 147–8.

which is distinguished from the others by this union of bodies. (2le3a2d, the first emphasis added)

This means that individuals under the attribute of extension are complex bodies, composed of many other bodies – in the end, of the simplest bodies. Furthermore, the identity of an individual depends on 'a certain fixed manner' in which bodies 'communicate their motions to each other'. What does Spinoza have in mind? The following lemma clarifies things:

If the parts composing an Individual become greater or less, but in such a proportion that they all keep *the same ratio of motion and rest to each other* as before, then the Individual will likewise retain its nature, as before, without any change of form. (2le5, emphasis added)

Many commentators have emphasized the central importance of the notion of a fixed relation (*ratio*) for Spinoza's theory of individuals, and rightly so. Obviously, the relation of the simplest bodies is decisive for the identity of composite extended entities. The importance of relationality can be seen as stemming from the field character of Spinoza's metaphysics of extension: distribution in the unified field of power is always relational and established by the way the intensity of power is dispersed *between* different spatial regions. I interpret this to mean that the metaphysical description of an individual involves a fixed arrangement of interaction between power intensifications, and that at the physical level of description this equals the relation of motion and rest between the bodies composing the individual.

Spinoza speaks of the structure (*fabrica*) of the human body in several places,[45] which suggests that complex bodies are concrete corporeal structures constituted by innumerable simpler bodies in the required interactive configuration. And although Spinoza suggests in the *Short Treatise* (*KV* IIpr; C, p. 96; G I, p. 52) that the relation characterizing an individual can be expressed as a mathematical ratio – such as 1 to 3 – holding between the sum of motion and the sum of rest of the individual's parts, I do not think that such a simple numerical proportion tells the whole story concerning corporeal structures.[46] I would suggest that the differences in physical entities' structures correspond to differences in their essences.[47] This is not an issue

[45] See 1app (C, p. 443; G II, p. 81), 3p2s (C, pp. 495–6; G II, pp. 142–3), 4p59s.

[46] For discussions on the ratio of motion and rest, see e.g. Jonas 1965, pp. 48–9; Deleuze 1992 (1968), p. 208; Matheron 1971, pp. 230–2; Rice 1971, p. 648; Bennett 1984, p. 232; Ablondi and Barbone 1994, p. 79; Garber 1994, p. 54; Garrett 1994, pp. 82–7; Barbone 2002, p. 98; Lin 2005, pp. 248–51; Della Rocca 2008b, pp. 148–50; Di Poppa 2010, pp. 286–7; LeBuffe 2010, pp. 103–4.

[47] A strong correspondence between natures, their powers, and structures seems to be made in 3p2s, where Spinoza signals that knowing 'what can be deduced from the consideration of its [the Body's]

Spinoza would discuss in great detail, but there is one revealing passage which links power with structure:

> The act of beating, insofar as it is considered physically, and insofar as we attend only to the fact that the man raises his arm, closes his fist, and moves his whole arm forcefully up and down, is *a virtue, which is conceived from the structure of the human Body* [*virtus est, quae ex Corporis humani fabrica concipitur*]. (4p59s, emphasis added)

As virtue equals power of acting (4d8), it can be said that Spinoza is here speaking about physical power producing effects according to the body's structure. This implies, I think, that the metaphysically fundamental striving power is expressed in different forms under different attributes; under extension, every particular physical thing can be conceived as a complex composition or structure of powerful states that cause physical effects without ceasing.

Now, as Nadler explains, the mode of explanation favoured by the early modern philosophers such as Descartes invokes structures of complex entities and can thus be called structural;[48] evidently, as far as physics is concerned, Spinoza endorses the basics of this line of thought. But it is also evident that Spinoza does not think that structure- and power-based explanations are in conflict but that there is a concordance between the ontologico-dynamistic and physico-structural aspects of things; indeed, they obviously are, for him, interconnected. All nature is intrinsically causally efficacious; all things have, in virtue of their essential power, a causal 'kick' that is manifested differently under different attributes: under extension as physical effects the nature of which depends to a great degree on how bodies are structured, and under thought as affirmations the nature of which depends on the complexity of the minds that do the affirming. Things are, in essence, strivers; as physical beings, structurized strivers.

The aforesaid has implications for what kind of theory of explanation is available for Spinoza. As a matter of fact, the present interpretation explains why Spinoza does not regard power-evoking explanations as vacuous or occult: they involve the structural element required by the mechanical mode of explanation. I think Des Chene is exactly right in contending, '[t]he problem with Molière's doctor is not that he appealed to a *virtus dormitiva* – which, after all, opium does have – but that the inquiry was supposed to end

nature alone' equals knowing 'what the body can do'; and we can succeed in the latter by discerning accurately 'the very structure of the human body' (C, p. 496; G II, p. 143; see also C, p. 495; G II, p. 142).

48 Nadler 1998, p. 520.

there'.[49] This is not what Spinoza does, because for him the structural features of physical things are explanatory; they just are not the metaphysically basic features of things. He would seem to think that talk about essences and their effects is of a general metaphysical level, and that those essences have different 'concrete manifestations' under God's attributes: bodies structured in certain causally efficacious ways;[50] minds of a differing complexity, some of them capable of affirming adequate ideas.[51] Thus, when power-based explanations are connected to these kinds of concretizations, reference to power is not vacuous.

Interestingly, Spinoza remarks that it does not matter which particular parts comprise a physical individual (for example a human body) at a given time:

> If, of a body, *or* of an Individual, which is composed of a number of bodies, some are removed, and at the same time as many others of the same nature take their place, the [NS: body, or the] Individual will retain its nature, as before, without any change of its form. (2le4)

As is well known, a great portion of any human body is renewed several times during its lifetime. Interpreted from the viewpoint of power, a human body is constituted by an extremely complex arrangement of power intensifications which must hold a certain fixed relation for the body's temporal existence to continue.

Here, in fact, we find a theory of *bodily* constitution at work: recall that constitution was designated as a focal notion in the Spinozistic account of change. Spinoza tells us that, for instance, the body is constituted differently when it is hungry and when it is replete (3p59s). It is evident that our bodily constitution constantly varies, for the human body requires 'many other bodies, by which it is, as it were, continually regenerated' (2po4). Under extension, a concrete bodily structure consisting of innumerable actual modifications may undergo change and still realize in temporality a certain ratio of motion and rest: Spinoza makes it clear that the components that constitute a thing can change as long as the essential arrangement prevails.[52] Apparently, he thinks that the idea of a certain ratio or mutual arrangement (of states) allows a kind of multiple realizability of an essence and refers to it as essence-as-constituted; as Lin notes, '[t]he plausibility of Spinoza's view requires that a person's *ratio* be construed broadly enough that it can be realized by a diversity of states, and so survive the changes that a person

[49] Des Chene 1996, p. 156. [50] See esp. 3p2s. [51] See the next section. [52] See 2le4–le7.

typically undergoes in the normal course of things'.[53] Evidently, it is critical that the relation or arrangement of states characteristic of a thing persists even when those states are partially produced by external causes. In fact, for a thing not to be at least partly 'constructed' out of externally caused affections it would have to stay in existence purely actively, unaffected by any external cause, and this is impossible for finite beings.[54] In other words, the actual being can involve both active and passive elements to a changing degree as long as those elements remain in a certain mutual relationship. As 2le7s states, 'a composite Individual can be affected in many ways, and still preserve its nature'.[55]

So, when an extended mode moves or its parts change (2le4–le7), this requires extremely rapid reorganization in the field of power – the regions the mode fills often change completely. To illustrate this, we can consider, for example, a tennis ball moving swiftly across a tennis court. The ball (i.e. a determinate physico-kinetic structure) is, metaphysically speaking, a spatial power intensification pattern. And from this it follows that as the ball moves the intensifications that constitute it occupy continually new regions of space, but in such a fashion that despite this continual spatial alteration they are maintained in the arrangement corresponding to the essence of the tennis ball. In other words, the unified power field alters so that subsequent regions are filled by the pattern of intensifications physically conceptualized as the ball.[56] Thus, the infinite spatiotemporal field of power is in a state of flux, ceaselessly reordered, while power as a whole stays the same. And there comes a time when our limited *conatus* power can no more oppose the decomposing external causes, and the structure of our body is no longer renewed.[57] As a result, no region in the power field corresponds to our essence – an event more familiarly called death (see 4p3, 4p39s). Of course, power that used to be arranged in a manner called 'a human being' does not disappear, it is simply redistributed anew.

[53] Lin 2005, p. 262. [54] See 4p3–p4.

[55] Matheron 1971 (pp. 230–2) and 1988 ([1969], p. 44) offer interesting discussions of these matters.

[56] For a recent paper that presents arguments for regarding Spinoza as a process ontologist, see Di Poppa 2010.

[57] To illustrate the role of *conatus* here, we may draw on Spinoza's analogy of a worm that lives in the blood. Spinoza writes that 'all the parts [of the blood] are *controlled* by the overall *nature* of the blood and compelled to mutual adaptation as the overall nature of the blood requires, so *as to agree with one another in a definite way*' (*Ep*32; S, p. 849; G IV, p. 171, emphases added). I believe this to mean that whatever parts the total 'package' that is the actual individual is composed of, they must 'agree' with each other in this way for the individual to survive; evidently, this agreement requires constant (and essence-originating) activity or power of 'controlling' that distinguishes an individual from a mere aggregate of finite things. Cf. Rice 1971, pp. 648, 654; Barbone 2002, pp. 96–7, 100.

CONATUS, THOUGHT, AND REASON

I have argued that the attribute of extension should be seen as a spatial field of power, and that physical individuals are, metaphysically speaking, patterns of power intensifications that occupy locations in that field. It is, however, much more difficult to say how the attribute of thought should be dynamically conceived.[58] Clearly, there is power of thinking that corresponds to physical power of acting; even more importantly, Spinoza held from early on that thoughts are mental acts of affirming and as such they are endowed with power: '[I]t [the will] is *a thought, i.e., a power of doing each one, of affirming and of denying*' (*CM* II.12; C, p. 346; G I, p. 280, emphasis added). The thesis that ideas are powerful affirmations has, I think, the following consequence with regard to Spinoza's monism as a whole: it is clear that, for Spinoza, all ideas are necessitated by God's nature; but since ideas are affirmations and affirming requires power, it follows, once again, that realizing the necessitarian system requires power.

The Spinozistic idea that all things are causally efficacious has, especially if we keep in mind that for him all things are animate in different degrees (2p13s), the obvious upshot that all things must be endowed with at least some power of thinking. In a recent paper, Garrett develops this line of thought and notes that 'all finite individuals are conscious to at least some degree'.[59] At first blush an incredible doctrine no doubt; but, according to Garrett, not one without its merits: as Spinoza identifies degrees of consciousness with degrees of power of thinking, many of the explanatory demands on his theory of consciousness can be met. The resulting picture Garrett dubs Spinoza's 'incremental naturalism' according to which there is no radical rupture between human intellectual life and the rest of nature: the former results when a large number of ideas of simpler bodies form a complex, unified whole endowed with a high level power of thinking.[60]

That ideas contain something intrinsically active is a tenet that echoes throughout the *Ethics*:

> By idea I understand a concept [*conceptum*] of the Mind that the Mind forms because it is a thinking thing.
>
> Exp.: *I say concept rather than perception, because the word perception seems to indicate that the Mind is acted on by the object. But concept seems to express an action of the Mind.* (2d3)

[58] For one good suggestion, see Rice 1999, p. 162. [59] Garrett 2008, p. 23.
[60] Ibid., esp. pp. 18–19, 23–5.

The view Spinoza explicitly denies is

> that an idea is something mute, like a picture on a tablet, and not a mode of thinking, viz. the very [act of] understanding. (2p43s)

And in the early *Ep*32 (S, pp. 849–50; G IV, pp. 173–4) Spinoza equates the human mind with a certain determination of God's infinite power of thinking. Thus, completely in keeping with what has been said about all things generally, the human mind is, in essence, power, in this case power of thinking that brings about ideas by affirming things.[61] When the causal history of the things affirmed also involves external causes, the resulting ideas are inadequate; when the affirmed is of our making, as it were, the results are adequate.

Moreover, as '[e]ach idea of each body, or of each singular thing which actually exists, necessarily involves an eternal and infinite essence of God' (2p45) and any human mind thereby has, on the basis of *any* idea of *any* thing 'an adequate knowledge of God's [. . .] essence' (2p47), it can be said that even *every inadequate* idea of any finite thing, or object, involves an *adequate* idea of its attribute. The reasoning behind this looks to be that there are concepts that are conceived or understood through themselves, namely those of attributes, and through them we can grasp finite things because those things are, in effect, modifications of attributes that are conceived through themselves. But as it is Spinoza's contention that we are adequate (i.e. complete) causes of our adequate ideas, it follows by 3d2 – and nicely in keeping with the claim that all our ideas are in some sense active – that this adequate element in our ideas must be *actively* brought forth by our minds. Here I agree with Della Rocca who notes that each idea must involve something active already by virtue of being an idea: '[A]n idea gets to be an affirmation in virtue of being an action.'[62] But as activity can only result in adequacy, there must be something (epistemically) adequate in each idea. Precisely here we encounter the main problem of this section: why would the free affirmations, or our mental activity, result in adequate, i.e. clear and distinct ideas? In alternative terms: given that our minds can be constituted in different ways, why would adequate thoughts follow from the self-determined constitution?

[61] I believe that my view is in harmony with Della Rocca's and Steinberg's important recent discussions concerning power and activity involved in thinking. Della Rocca (2003b) contends: '[E]ach idea is inherently an *affirmation* that something is the case' (p. 202); 'in effect, the mind does nothing but affirm things' (p. 206). According to Steinberg (2005, p. 150), ideas are 'dynamic entities', and '[t]he strength of an idea is [. . .] shown in the fact that it generates other new ideas and exercises a general influence over the content of our thought' (pp. 151–2).

[62] Della Rocca 2003b, p. 207.

The problem becomes even more pressing when the *conatus* doctrine is taken into account. It is a central tenet of Spinoza's ethics that the intrinsic human striving is manifested, if not prevented by external causes and passions brought about by them, as *reason*. Indeed, this is the path that allows Spinoza to make the transition from rather 'blind' and egoistic-seeming striving to behaviour that most of us would acknowledge as more or less ethical – or at least highly intellectual, as is witnessed by the propositions stating that the 'highest virtue', i.e. power, of the mind 'is to know God' (4p28) and that people powerful enough to understand endeavour to bring it about that other people, too, would live under the guidance of reason (4p37, d). So, as Charles Ramond rightly observes, '[o]ne of the originalities of the [Spinozistic] system is, in fact, the constant linkage of "power" to "reason"'.[63]

As we have seen, the ingredient in Spinoza's geometry of power that underlies his thinking on causality, *conatus*, and power may be called dynamic essentialism, according to which things can do only what follows or flows from their nature. Now, it is important to note that this kind of dynamism does not, in itself, entail rationalism. Indeed, in the correspondence with Blyenbergh, Spinoza acknowledges this himself. Blyenbergh poses the following question:

Here again a question can be raised; if there were a mind to whose particular nature the pursuit of pleasure or villainy was not repugnant but agreeable, could he have any virtuous motive that must move him to do good and avoid evil? But how is it possible that one should be able to relinquish the desire for pleasure when this desire at that time pertains to his essence[?] (*Ep*22; S, pp. 830–1; G IV, pp. 141–2)

Spinoza's answer is interesting enough to be quoted in its entirety:

Finally, as to your third question, it presupposes a contradiction. It is just as if someone were to ask me whether, if it accorded better with a man's nature that he should hang himself, there would be any reason why he should not hang himself. However, suppose it possible that there could be such a nature. Then I say (whether I grant free will or not) that if anyone sees that he can live better on the gallows than at his own table, he would be very foolish not to go and hang himself. And he who saw clearly that he would in fact enjoy a more perfect and better life or essence by engaging in villainy than by pursuing virtue would also be a fool if he did not do just that. For in relation to such a perverted human nature, villainy would be virtue. (*Ep*23; S, p. 834; G IV, pp. 151–2)

But surely, the question stands: why does unhindered, essence-originating human striving manifest itself as reason in the sense of producing adequate

[63] Ramond 1999, p. 63, translation mine.

ideas only? Why are 'perverted' human natures, with an internal thrust to delinquency, for example, impossible?

I know of no detailed attempt to solve this problem, despite the fact that what is here at stake is the compatibility of Spinoza's two deepest commitments: rationalism and dynamism. The connection between these two tenets seems often to be regarded as quite unproblematic. But precisely this connection needs to be proved: on what grounds can Spinoza maintain that activity – or 'acting according to the laws of our mind's nature alone' – equals rationality? As becomes clear as soon as one reflects on it, 'striving' and 'activity' have no necessary conceptual linkage with 'rationality' and 'truth'. Friedrich Nietzsche, a dynamistic thinker in his own right, expresses this point with his characteristic flare in *Beyond Good and Evil*: 'The will to truth[.] [. . .] We asked about the *value* of this will. Granted, we will truth: *why not untruth instead*? And uncertainty? Even ignorance?' (*BGE* 1.1). Indeed, Nietzsche himself attributes surprising life-preserving capacities to error and untruth:

> We do not consider the falsity of a judgment as itself an objection to a judgment; this is perhaps where our new language will sound most foreign. The question is how far the judgment promotes and preserves life [. . .] And we are fundamentally inclined to claim that the falsest judgments [. . .] are the most indispensable to us [. . .] – that a renunciation of false judgments would be a renunciation of life, a negation of life. To acknowledge untruth as a condition of life: this clearly means resisting the usual value feelings in a dangerous manner. (*BGE* 1.4)

To my mind, in this passage we encounter a thinker who is not too far removed from Spinoza in that for him the doctrine of will to power as life-enhancement is the fundamental point of departure. This, however, does not keep Nietzsche from holding that our fundamental striving to live and to expand our power is linked up with, and even presupposes, untruth instead of truth and adequacy. Here we can see a challenge emerging – and an especially acute one for anyone who makes the striving to persevere in being the very essence of all things. In Spinozistic terms, the central question concerns the notion of adequacy: on what grounds is Spinoza allowed to hold that causal adequacy (i.e. being the only and whole cause of an effect in virtue of one's essence) results in epistemic adequacy (i.e. clear and distinct ideas)? Why are adequately, i.e. actively, produced ideas clear and distinct? Or in terms of the *conatus* doctrine: why does our striving, to the extent it is free, amount to reasoning?

Descartes faced a similar problem concerning the special epistemic status of innate ideas he found present within himself: obviously, for him, only these ideas are clear and distinct and, consequently, beyond doubt. And just as we can ask Spinoza why should intrinsic power produce adequate ideas, we can ask Descartes why should innateness be necessarily linked with truth. As is well known, Descartes takes pains to show that God exists, and he thinks that the Cartesian God provides an escape from these problems. In the Fourth Meditation he argues as follows:

> [E]very clear and distinct perception is undoubtedly something, and hence cannot come from nothing, but must necessarily have God for its author. Its author, I say, is God, who is supremely perfect, and who cannot be a deceiver on pain of contradiction; hence the perception is undoubtedly true. (CSM II, p. 43)

In other words, because we can put our trust in a veracious God who has imprinted certain ideas on our minds, those innate ideas are necessarily true. Regardless whether we find Descartes's thinking here convincing or not, Spinoza's overall metaphysical position, especially his immanent and de-anthropomorphized conception of God, seems to be so unlike the Cartesian one that this kind of defence is hardly available for him. So, the question is: what kind would be?

The following argument is probably the most obvious one; as I understand them, such scholars as Parkinson and Bennett endorse at least some variant of it,[64] and I suspect that it is taken for granted in many readings of Spinoza. According to Spinoza, that a finite thing x is active or adequate cause means that it is the sole, i.e. full or complete, cause of an effect (3d1, 3d2). Now it can be claimed that Spinoza implicitly assumes that being the joint product of two or more disparate things is the only possible source of inadequacy,[65] and since, as we saw above, he thinks that nothing that follows from x's activity alone is a joint product, it cannot be inadequate. Thus, acting adequately generates only ideas that are clear and distinct.

This is a valid argument, but its soundness surely raises doubts. Why should being the joint product of two or more things be the only source of inadequacy? Even if we admitted that an idea or mental state produced by both x and y is necessarily inadequate, this would not exclude the possibility of an idea resulting only from x's activity that might be inadequate on some other grounds. Moreover, this line of argument provides no positive explanation of what the adequacy of actively produced ideas could rest on. But, leaving these considerations aside, in the end everything hinges on the implicit premise stating that only joint causing results in inadequacy: if

[64] See especially Parkinson 1981, pp. 15–16; Bennett 1984, p. 178.

[65] According to Spinoza, any idea a finite thing x derives from sense-experience is a joint product of x's nature and external causes and therefore confused, that is, inadequate (2p16, 2p25, 2p28).

we could clarify Spinoza's reasons for holding it and formulate a convincing argument for it, we would have made considerable progress towards understanding the linkage between causal and epistemic adequacy; indeed, discerning Spinoza's reasons for this idea is precisely what I will later endeavour to provide. So, as it stands, this argument is unsatisfactory.

An interesting way to defend Spinoza can be framed from a very different angle, based on Spinoza's ethical views as discussed by Koistinen.[66] Although Koistinen himself neither presents nor defends the following argument, something akin to it can, I think, be extracted from what he says about the good and pleasurable in Spinoza's ethics, and it runs as follows. Joy is an emotion constituted by an increase in a human being's power of acting (3p11s), and since Spinoza holds that we strive, as far as we can, to increase our power of acting (3p12), it follows that we strive to do things that make us joyous. Now, Spinoza thinks that understanding is the only thing that gives us always nothing but joy, never sorrow (4p27 says that '[w]e know nothing to be certainly good or evil, except what really leads to understanding or what can prevent us from understanding'), and therefore it can be claimed that we strive, as far as we can, to understand. In other words, our free striving leads to adequate thinking, for it is the only exclusively joy-giving thing.[67]

This line of argument certainly has much to recommend it: the notions of power, joy, striving, and understanding are undeniably closely connected in the *Ethics*. In the Spinozistic framework, we are essentially joy-desiring and sorrow-averting beings, and since, for Spinoza, understanding is a joyous thing, he most probably thinks us to be thrust, in virtue of our essence, towards adequate thinking. The difficult question is, however, how do we obtain the idea that only adequate thinking is a certainly joy-giving activity? It should be noted that the aforementioned 4p27 is based on 4p26 and its claim that reasoning follows from our mind's essence and informs us about the goodness, and hence joyfulness, of understanding – but this, of course, already assumes what we are trying to prove, namely that adequate mental causation equals forming adequate ideas. Spinoza simply does not tell us why reasoning, understanding, or becoming more perfect is a joyful thing.

One way to build a non-question-begging defence of the joyfulness of understanding is to invoke our experience of how things affect us and the inductive inferences made on the basis of these experiences; all this could

[66] Koistinen, forthcoming$_b$.

[67] Also Christian Lazzeri (1999, p. 25) points towards a similar line of interpretation.

allow us to conclude that understanding is the only truly good, i.e. exclusively joy-giving, thing there is. In other words, human *conatus* would *eventually* lead to understanding. Actually, to think that 'thus far, understanding is the only thing that has never hurt us, but has only helped us, so let's strive to understand' sounds like no bad idea; but surely, this line of reasoning does not fulfil Spinoza's requirements of rigour and ideal of watertight deductive derivations from sound definitions and indubitable axioms. Thus, it does not provide us with a properly Spinozistic way of tying the two notions of adequacy together, and so it is only understandable that in 4p26 and 4p27 Spinoza does not rely on it.

I believe that the following argument that turns on God's omniscience in fact reveals Spinoza's cast of mind on these matters. Although this line of thought is by no means hidden, I have encountered no explicit analysis of it in the literature. First, it should be remembered that according to Spinoza, God has an adequate idea of everything there is (1p16, 2p3, 2p11c, 3p1d). Second, that a finite thing x is an active or adequate cause means that it is the sole, i.e. full or total, cause of an effect (3d1, 3d2). Third, it is a basic Spinozistic tenet that knowledge of an effect requires knowledge of the cause (1a4), and this together with the aforesaid implies that when x acts, God has an idea of its effect that is formed solely through x. Thus, for God's idea of that effect to be (epistemically) adequate the idea actively (i.e. causally adequately) caused by thinking x must be (epistemically) adequate. In other words, unhindered mental striving, or active thinking, must produce adequate ideas only, for otherwise all God's ideas would not be adequate, and this would violate his omniscience because when x acts, God has some of his adequate ideas only insofar as he is expressed as x. To my mind, this is the reason why Spinoza sees no problem with his two notions of adequacy, epistemic and causal. That this really is Spinoza's view on these matters is corroborated by the fact that the aforementioned 4p27, presenting the idea of understanding as the only true good, can be traced back to 2p40; and according to its demonstration,

> when we say that an idea in the human Mind follows from ideas that are adequate in it, we are saying nothing but that (by p11c) in the Divine intellect there is an idea of which God is the cause, not insofar as he is infinite, nor insofar as he is affected with the ideas of a great many singular things, but insofar as he constitutes only the essence of the human Mind [NS: and therefore, it must be adequate]. (2p40d)

It is quite natural, by my lights at least, to take Spinoza's way of thinking here to be resting on the just presented argument.

Of course, this last argument is only as strong as its premises are. Obviously, the ingredient containing the doctrine of God's omniscience is by far the most dubious one; I suspect that even many of those sympathetic towards other parts of the argument are not able to accept it as it stands. Be that as it may, the argument's precise meaning and Spinoza's reasons for holding it are worth examining. In the end, the question of God's omniscience boils down to the all-important 1p16 and its idea that because God produces everything he understands, the causal and epistemic realms fall into line; there can be no gap between causing and understanding, which makes all God's ideas adequate. The Spinozistic God turns out to be, as it were, the ultimate constructivist: there could never be anything outside the immanent God that could give rise to disagreement between God's intellect and things produced by God. Here we encounter, just as we did in Chapter 1, Spinoza's variant of the age-old Platonic vision of reality as an eternal and totally intelligible whole, which has, of course, been largely discarded. This Spinozistic (i.e. necessitarian and anti-teleological) version of it can be formulated in more modern terms as a doctrine according to which there is all-producing power-nature and a complete description of everything that ever obtains existence;[68] maybe this kind of de-theologized metaphysic fares better in our eyes than some traditional ones, but we are still surely not willing simply to accept it as evident. However, I think that a couple of things can be learned from all this. First, as with Descartes, Spinoza's answer to the question concerning the relation between innateness and truth has to do with the conception of God, only this time we are dealing with a non-anthropomorphic God, i.e. nature. Second, Spinoza's dynamic rationalism does not simply overlook the problem pertaining to the two notions of adequacy; on the contrary, the idea that we are rational to the extent we realize what follows from our essence is deeply rooted in Spinoza's ontological commitments, and revealing this sheds some new light on them.

The claim that unhindered *conatus* results in rationality goes well together with Spinoza's view according to which, as I see it, persevering in being cannot be reduced to mere survival or prolongation of temporal existence. This allows him to build an ethic that, although based on a basically egoistic principle – which has certain key aspects in common with the Hobbesian one – ends up not only declaring such noble-sounding propositions as '[h]ate can never be good' (4p45) and '[t]he good which everyone who seeks virtue wants for himself, he also desires for other men' (4p37) but also championing the renowned intellectual love of God that arises from adequate understanding (5p32–p37).

[68] I owe this formulation to Koistinen.

All this clearly implies that finding a connection between our conative character and adequate cognition is by no means of minor importance for Spinoza, for especially the intellectualism of the last part of his *Ethics* requires such a connection to exist. And what I hope to have shown is that, according to Spinoza, when we are acting autonomously, we are rational precisely *because* we are parts of an intelligible totality. Seen in a wider context, during the seventeenth century the traditional idea of the universe as a teleologically ordered rational system was threatened by the mechanistic new science, and therefore Spinoza's project can also be interpreted as an attempt to find a place for the traditional conviction in the newly adopted anti-teleological world-view. So, if Spinoza's reasoning – despite its naturalism and anti-anthropomorphism – that relies on God's omniscience does not appeal to us, it only goes to show that he, like so many others, was fighting a losing battle: having lost confidence in the world as an intelligible whole, we are inclined to call into question the linkage between power and reason. In this respect our sentiments are still echoing the intellectual cataclysm that took place in nineteenth-century thought, to a great extent due to the upheaval dynamistic metaphysics underwent in the hands of another modern monist, the disillusioned Schopenhauer.[69]

[69] For more on this, see my 2009a, esp. pp. 328–9.

Conclusion

From the preceding chapters emerges a philosophical position in which power and related notions have such a prominent place that it is fully justified, I think, to call Spinoza's metaphysics dynamistic. To conclude, I would like to make some general observations about this metaphysics. We can begin by noting how such notions as striving, power of acting, desire, and essence work together to form an elaborate system of concepts with which Spinoza can theorize about human existence. For Spinoza, power is first and foremost a metaphysical concept that refers to things' intrinsic capacity to cause effects; most importantly, the necessary realization of everything that follows from God's essence requires power. As finite things express this power, they enter into causal relations with each other and are essential causers that never fail to bring about effects by their intrinsic power.

The notion of power had fallen to disrepute in Spinoza's time, but Spinoza is determined to show that this was for no good reason. According to him, there is nothing occult in the intrinsic power things are endowed with; on the contrary, that power is exercised in the most intelligible way we can think of, with the kind of necessity that reigns in geometry. This is so because each thing has an internal structure most easily discernible in geometrical objects; that there is a certain register of variation an actual existent can undergo when under the varying influence of external causes – determined from eternity – does nothing to diminish the intelligibility and necessity involved. Indeed, as I hope to have shown, it is precisely on this that Spinoza founds his theory of human emotions designed to proceed in geometrical manner.

Strikingly, Spinoza offers us even more than this. First, he develops an elaborate theory according to which a specific kind of power individuates a thing: as long as God's power is modified or arranged according to an individual essence (so that the thing 'expresses the power of God' and there is power 'conceived from' the thing's structure), there is a thing continuing its existence with 'the same power' even though the total amount, strength, or intensity of this power varies. So, Spinoza displays a remarkable sensitivity to the contentions of the Eleatic Stranger by developing the most systematic view of individual existence as realization of being through striving power.

Second, appreciating Spinoza's dynamistic tendencies allows us to discern a basically non-teleological theory of human action in which we are considered to be, in essence, powers that take the path of least resistance; such a theory is supposed to meet the challenge of advancing a fully unified theory of action that Spinoza's naturalism requires. What sets cognitive agents apart from non-cognitive ones is the fact that they have ideas whose representational content is causally potent through having an effect on how the agent's striving is directed by emotions, i.e. ideas that indicate the encountered resistance. Moreover, when our striving is determined – irrespective of how this happens – to something specific, we can be said to desire something; as Spinoza makes clear, we often have conflicting desires, the strongest one prevailing over the others and causing its effect, i.e. leading to a specific action. Having established this kind of basic theory of human action, Spinoza confidently sets out to elaborate his naturalistic ethics the culminating point of which is formed by the contention, having its source in the unwavering confidence in the world being a completely rational whole, that unhindered use of our intrinsic power can only lead to rationality, joy, and peace of mind.

Despite being so unmistakably original, Spinoza's thought firmly connects with the pre-modern philosophical tradition. This holds especially with regard to his essentialism which is, due to rehabilitating the connection between essences and power, in an important respect more traditional than the ones espoused by Descartes and Hobbes. In Spinoza's system, individual essences occupy a place quite similar to the one possessed earlier by substantial forms in Aristotelianism; both are focal ingredients in serious attempts to solve those most profound metaphysical problems that pertain to individuation and causation. But of course, Spinoza does not just take some doctrine handed down by the tradition and make it a part of his monistic framework; the essentialism he offers is changed from the roots. Final causes are missing from Spinoza's world whose structure is modelled after geometry. Nothing is guided or governed by ends, but there is still order and stability in the world, just as there is in geometry. Things act according to their essences, but precisely as little finality is involved here as is in a triangle's essence determining the internal structure and properties of the figure. The challenge, as we have seen, is to apply this basic insight into the realm of temporal entities that undergo change. The new physics, in turn, has a profound impact on Spinoza's way of understanding intrinsic power as striving: the distinction between potentiality and actuality is discarded, the exercise of power has no natural termini, and striving power is seen as something resilient enough to offer the required grounding not only for inertial motion but also for resistance to opposition. Very excitingly, Spinoza fuses old and new together to form an ingenious dynamistic metaphysic which, while lacking many of those intricacies that had been seen to plague the scholastic theories, applies to human existence as well as to interactions taking place between bodies.

The theory presented here is no doubt compelling. But the difficult – in the end perhaps the most pertinent – question is, to what extent is it acceptable for us? Is it just a fascinating metaphysical fairy-tale, simply unbelievable regardless of its consistency? I would like to make some suggestions as to how these questions might be approached. Now, it seems to me that Spinoza's monism contains elements which we surely find baffling, such as the claim that everything that exists follows from God's essence, and with the same kind of necessity as in geometry. Given the centrality of these contentions, is Spinoza's system permanently out of our reach, doomed to the status of a mere curiosity? I would say *no*, especially for the following reason: it seems to me that it is possible to formulate a rough approximation of the metaphysics of the opening part of the *Ethics* in terms that make it considerably less foreign to our contemporary eye. If one is prepared, as I suppose many of us to be, to entertain the ideas that, first, nature is a whole beyond which there is nothing, second, that this nature as well as finite entities as its parts are intrinsically powerful, and third, that certain features are more constitutive (i.e. essential) to things than others, enough is in place to go on reading and taking quite seriously the rest of the *Ethics*, especially its third and fourth parts. That its brand of naturalism still appeals to us is also witnessed by the many similarities Spinoza's dynamic essentialism bears to certain notable contemporary positions in metaphysics and philosophy of science.

It is, however, its connection to the vision of good life that most contributes to making Spinoza's geometry of power so inspiring and unique: in it power is inextricably linked to the good, the latter being what increases our power, happiness equalling having a maximally high level of power of acting. In the final analysis, freedom and virtue are about unhindered realization of our power. Thus Spinoza's ethical perfectionism is nicely backed up by – indeed, probably impossible without – his dynamistic theory according to which we strive to exist so that we are determined by our own essence alone. And here also Spinoza is very far from being of merely antiquarian interest: ever since the Romantics and Nietzsche, the view of humans as beings who strive, with all their power, for their own particular kind of self-realization is one that hardly fails to strike a chord with us.

The *Ethics* is a notoriously difficult and highly abstract work, but Spinoza clearly thinks that many of its lessons are firmly anchored in our life experience. And indeed, Spinoza's dynamism can readily explicate what happens when someone who has recently lost a loved one cannot do anything but just 'hold on' to existence or when a creative person experiences a high level of self-fulfilment. Moreover, Spinoza holds that there is a conscious form of our essential striving: desire; so through the desires we are so familiar with, it is possible to gain experiential access to our dynamic nature. It is, however, one thing to find dynamic terms somehow fitting when attempting to describe the things we undergo and quite another to elaborate a comprehensive dynamistic theory of the human condition; the latter is what Spinoza sets out to offer and, to an astonishing degree by any standards, delivers. One way to see Spinoza's ethical position is perhaps this: our power is necessarily limited and under opposition, which is exactly the reason why it should be used as wisely as possible, as explicated in the *Ethics*. We always desire something, but it matters what we desire, for some desires are more rational and active than others. Reason and affectivity are tightly intertwined in Spinoza's philosophy, because they both have their basis in the essential *conatus*. Accordingly, rationality does not imply lack of emotions, for strictly speaking we are never desireless. Granted, human existence is a fragile affair of innumerable conflicts, sorrows, and tumults of emotion; but the comforting message Spinoza wants to convey is that there is a way to attain, not an unemotional existence without struggles, but a joyful and fulfilling life of the Spinozistic sage, the rationally desiring being.

Bibliography

Abbruzzese, John Edward. 2007. 'A Reply to Cunning on the Nature of True and Immutable Natures'. *British Journal for the History of Philosophy* 15 (1): 155–67.

Ablondi, Frederick and Steven Barbone. 1994. 'Individual Identity in Descartes and Spinoza'. *Studia Spinozana* 10: 69–91.

Allison, Henry E. 1987. *Benedict de Spinoza: An Introduction*. Revised edition. New Haven: Yale University Press.

Barbone, Steven. 2002. 'What Counts as an Individual for Spinoza?' In *Spinoza: Metaphysical Themes*, ed. Olli Koistinen and John Biro, pp. 89–112. Oxford University Press.

Barbone, Steven and Lee Rice. 1994. 'Spinoza and the Problem of Suicide'. *International Philosophical Quarterly* 34 (2): 229–41.

Bennett, Jonathan. 1983. 'Teleology and Spinoza's Conatus'. *Midwest Studies in Philosophy* 8 (1): 143–60.

1984. *A Study of Spinoza's* Ethics. Cambridge University Press.

1990. 'Spinoza and Teleology: A Reply to Curley'. In *Spinoza: Issues and Directions. The Proceedings of the Chicago Spinoza Conference*, ed. Edwin Curley and Pierre-François Moreau, pp. 53–7. Leiden: E. J. Brill.

1996. 'Spinoza's Metaphysics'. In *The Cambridge Companion to Spinoza*, ed. Don Garrett, pp. 61–88. Cambridge University Press.

2001. *Learning from Six Philosophers: Descartes, Spinoza, Leibniz, Locke, Berkeley, Hume. Volume I*. Oxford: Clarendon Press.

Bove, Laurent. 1996. *La stratégie du conatus. Affirmation et résistance chez Spinoza*. Paris: Vrin.

Carraud, Vincent. 2002. *Causa sive ratio. La raison de la cause, de Suarez à Leibniz*. Paris: Presses Universitaires de France.

Carriero, John. 1991. 'Spinoza's Views on Necessity in Historical Perspective'. *Philosophical Topics* 19 (1): 47–96.

1994. 'On the Theological Roots of Spinoza's Argument for Monism'. *Faith and Philosophy* 11 (4): 626–44.

1995. 'On the Relationship between Mode and Substance in Spinoza's Metaphysics'. *Journal of the History of Philosophy* 33 (2): 245–73.

2005. 'Spinoza on Final Causality'. In *Oxford Studies in Early Modern Philosophy, Volume II*, ed. Daniel Garber and Steven Nadler, pp. 105–47. Oxford: Clarendon Press.

2009. *Between Two Worlds: A Reading of Descartes's* Meditations. Princeton University Press.

Castañeda, Hector-Neri. 1975. 'Individuation and Non-Identity: A New Look'. *American Philosophical Quarterly* 12 (2): 131–40.

Chalmers, Alan F. 1999. *What Is This Thing Called Science?* Third edition. Buckingham: Open University Press.

Chappell, Vere. 2005. 'Self-Determination'. In *Early Modern Philosophy: Mind, Matter, and Metaphysics*, ed. Christia Mercer and Eileen O'Neill, pp. 127–41. Oxford University Press.

Cohen, I. Bernard. 1964. '"Quantum in se est": Newton's Concept of Inertia in Relation to Descartes and Lucretius'. *Notes and Records of the Royal Society of London* 19 (2): 131–55.

Cunning, David. 2003. 'True and Immutable Natures and Epistemic Progress in Descartes's *Meditations*'. *British Journal for the History of Philosophy* 11 (2): 235–48.

Curley, Edwin. 1969. *Spinoza's Metaphysics: An Essay in Interpretation*. Cambridge, MA: Harvard University Press.

1988. *Behind the Geometrical Method: A Reading of Spinoza's* Ethics. Princeton University Press.

1990. 'On Bennett's Spinoza: The Issue of Teleology'. In *Spinoza: Issues and Directions. The Proceedings of the Chicago Spinoza Conference*, ed. Edwin Curley and Pierre-François Moreau, pp. 39–52. Leiden: E. J. Brill.

Curley, Edwin and Gregory Walski. 1999. 'Spinoza's Necessitarianism Reconsidered'. In *New Essays on the Rationalists*, ed. Rocco J. Gennaro and Charles Huenemann, pp. 241–62. New York: Oxford University Press.

Delbos, Victor. 1950. *Le Spinozisme. Cours professé à la Sorbonne en 1912–1913*. Troisième édition. Paris: Vrin.

Deleuze, Gilles. 1992 (1968). *Expressionism in Philosophy: Spinoza*. Trans. Martin Joughin. New York: Zone Books.

1988 (1981). *Spinoza: Practical Philosophy*. Trans. Robert Hurley. San Francisco: City Lights Books.

Della Rocca, Michael. 1996. 'Spinoza's Metaphysical Psychology'. In *The Cambridge Companion to Spinoza*, ed. Don Garrett, pp. 192–266. Cambridge University Press.

1999. '"If a Body Meet a Body": Descartes on Body-Body Causation'. In *New Essays on the Rationalists*, ed. Rocco J. Gennaro and Charles Huenemann, pp. 48–81. New York: Oxford University Press.

2003a. 'A Rationalist Manifesto: Spinoza and the Principle of Sufficient Reason'. *Philosophical Topics* 31: 75–93.

2003b. 'The Power of an Idea: Spinoza's Critique of Pure Will'. *Noûs* 37 (2): 200–31.

2008a. 'Rationalism Run Amok: Representation and the Reality of Emotions in Spinoza'. In *Interpreting Spinoza: Critical Essays*, ed. Charlie Huenemann, pp. 26–52. Cambridge University Press.

2008b. *Spinoza*. London and New York: Routledge.

Des Chene, Dennis. 1996. *Physiologia: Natural Philosophy in Late Aristotelian and Cartesian Thought.* Ithaca: Cornell University Press.

Deveaux, Sherry. 2003. 'The Divine Essence and the Conception of God in Spinoza'. *Synthese* 135 (3): 329–38.

Di Poppa, Francesca. 2010. 'Spinoza and Process Ontology'. *The Southern Journal of Philosophy* 48 (3): 272–94.

Donagan, Alan. 1979 (1973). 'Spinoza's Proof of Immortality'. In *Spinoza: A Collection of Critical Essays*, ed. Marjorie Grene, pp. 241–58. Notre Dame: University of Notre Dame Press.

1988. *Spinoza.* Hemel Hempstead: Harvester Wheatsheaf.

Doney, Willis. 2005. 'True and Immutable Natures'. *British Journal for the History of Philosophy* 13 (1): 131–7.

Edwards, Jeffrey. 2000. *Substance, Force, and the Possibility of Knowledge: On Kant's Philosophy of Material Nature.* Berkeley: University of California Press.

Ellis, Brian. 2002. *The Philosophy of Nature: A Guide to the New Essentialism.* Chesham: Acumen.

Emilsson, Eyjólfur Kjalar. 2009. 'Plotinus on Act and Power'. In *The World as Active Power: Studies in the History of European Reason*, ed. Juhani Pietarinen and Valtteri Viljanen, pp. 71–87. Leiden: E. J. Brill.

Esfeld, Michael. 2002. *Holismus in der Philosophie des Geistes und in der Philosophie der Physik.* Frankfurt am Main: Suhrkamp.

Feingold, Mordechai. 2004. *The Newtonian Moment: Isaac Newton and the Making of Modern Culture.* New York: Oxford University Press.

Forsyth, T. M. 1974 (1927). 'Spinoza's Doctrine of God in Relation to His Conception of Causality'. In *Studies in Spinoza: Critical and Interpretive Essays*, ed. S. Paul Kashap, pp. 3–15. Berkeley: University of California Press.

Freddoso, Alfred J. 1988. 'Medieval Causation and the Case against Secondary Causation in Nature'. In *Divine and Human Action: Essays in the Metaphysics of Theism*, ed. Thomas V. Morris, pp. 74–118. Ithaca: Cornell University Press.

Gabbey, Alan. 1980. 'Force and Inertia in the Seventeenth Century: Descartes and Newton'. In *Descartes: Philosophy, Mathematics and Physics*, ed. Stephen Gaukroger, pp. 230–320. Brighton: Harvester Press.

Garber, Daniel. 1992a. *Descartes' Metaphysical Physics.* University of Chicago Press.

1992b. 'Descartes' Physics'. In *The Cambridge Companion to Descartes*, ed. John Cottingham, pp. 286–334. Cambridge University Press.

1994. 'Descartes and Spinoza on Persistence and Conatus'. *Studia Spinozana* 10: 43–67.

2001. *Descartes Embodied: Reading Cartesian Philosophy through Cartesian Science.* Cambridge University Press.

Garrett, Don. 1979. 'Spinoza's "Ontological" Argument'. *The Philosophical Review* 88 (2): 198–223.

1990. '"A Free Man Always Acts Honestly, Not Deceptively": Freedom and the Good in Spinoza's *Ethics*'. In *Spinoza: Issues and Directions. The Proceedings of the Chicago Spinoza Conference*, ed. Edwin Curley and Pierre-François Moreau, pp. 221–38. Leiden: E. J. Brill.

1991. 'Spinoza's Necessitarianism'. In *God and Nature: Spinoza's Metaphysics*, ed. Yirmiyahu Yovel, pp. 191–218. Leiden: E. J. Brill.

1994. 'Spinoza's Theory of Metaphysical Individuation'. In *Individuation and Identity in Early Modern Philosophy: Descartes to Kant*, ed. Kenneth F. Barber and Jorge J. E. Gracia, pp. 73–101. Albany: State University of New York Press.

1996. 'Spinoza's Ethical Theory'. In *The Cambridge Companion to Spinoza*, ed. Don Garrett, pp. 267–314. Cambridge: Cambridge University Press.

1999. 'Teleology in Spinoza and Early Modern Rationalism'. In *New Essays on the Rationalists*, ed. Rocco J. Gennaro and Charles Huenemann, pp. 310–35. New York: Oxford University Press.

2002. 'Spinoza's *Conatus* Argument'. In *Spinoza: Metaphysical Themes*, ed. Olli Koistinen and John Biro, pp. 127–58. Oxford University Press.

2008. 'Representation and Consciousness in Spinoza's Naturalistic Theory of the Imagination'. In *Interpreting Spinoza: Critical Essays*, ed. Charlie Huenemann, pp. 4–25. Cambridge University Press.

2009. 'Spinoza on the Essence of the Human Body and the Part of the Mind That is Eternal'. In *The Cambridge Companion to Spinoza's* Ethics, ed. Olli Koistinen, pp. 284–302. Cambridge University Press.

Gaukroger, Stephen. 2006. *The Emergence of a Scientific Culture: Science and the Shaping of Modernity, 1210–1685*. Oxford: Clarendon Press.

Gilead, Amihud. 1990. 'Spinoza's Two Causal Chains'. *Kant-Studien* 81 (4): 454–75.

Green, Leslie. 1998. 'Power'. In *Routledge Encyclopedia of Philosophy, Volume VII*, ed. Edward Craig, pp. 610–13. London: Routledge.

Gueroult, Martial. 1968. *Spinoza I. Dieu (*Éthique, *I)*. Paris: Aubier-Montaigne.

1974. *Spinoza II. L'Âme (*Éthique, *II)*. Paris: Aubier-Montaigne.

1980. 'The Metaphysics and Physics of Forces in Descartes'. In *Descartes: Philosophy, Mathematics and Physics*, ed. Stephen Gaukroger, pp. 196–229. Brighton: Harvester Press.

Harré, Rom. 1970. 'Powers'. *British Journal for the Philosophy of Science* 21 (1): 81–101.

Harré, Rom and E. H. Madden. 1973. 'Natural Powers and Powerful Natures'. *Philosophy* 48 (185): 209–30.

1975. *Causal Powers: A Theory of Natural Necessity*. Oxford: Basil Blackwell.

Harris, Errol E. 1972. 'Comment'. *The Canadian Journal of Philosophy* 2 (2): 191–7.

1973. *Salvation from Despair: A Reappraisal of Spinoza's Philosophy*. The Hague: Martinus Nijhoff.

Hatfield, Gary C. 1979. 'Force (God) in Descartes' Physics'. *Studies in History and Philosophy of Science* 10 (2): 113–40.

Hattab, Helen. 2003. 'Conflicting Causalities: The Jesuits, their Opponents, and Descartes on the Causality of the Efficient Cause'. In *Oxford Studies in Early*

Modern Philosophy, Volume I, ed. Daniel Garber and Steven Nadler, pp. 1–22. Oxford: Clarendon Press.

2007. 'Concurrence or Divergence? Reconciling Descartes's Physics with his Metaphysics'. *Journal of the History of Philosophy* 45 (1): 49–78.

2009. *Descartes on Forms and Mechanisms.* Cambridge University Press.

Hempel, Carl G. and Paul Oppenheim. 1948. 'Studies in the Logic of Explanation'. *Philosophy of Science* 15 (2): 135–75.

Hoffman, Paul. 2009. 'Passion and Motion in the New Mechanics'. In *Essays on Descartes*, pp. 125–41. Oxford University Press.

Huenemann, Charlie. 1999. 'The Necessity of Finite Modes and Geometrical Containment in Spinoza's Metaphysics'. In *New Essays on the Rationalists*, ed. Rocco J. Gennaro and Charles Huenemann, pp. 224–40. New York: Oxford University Press.

2004. 'Spinoza and Prime Matter'. *Journal of the History of Philosophy* 42 (1): 21–32.

2008. 'Epistemic Autonomy in Spinoza'. In *Interpreting Spinoza: Critical Essays*, ed. Charlie Huenemann, pp. 94–110. Cambridge University Press.

Hurka, Thomas. 1993. *Perfectionism.* Oxford University Press.

James, Susan. 1993. 'Spinoza the Stoic'. In *The Rise of Modern Philosophy: The Tension between the New and Traditional Philosophies from Machiavelli to Leibniz*, ed. Tom Sorell, pp. 289–316. Oxford: Clarendon Press.

1997. *Passion and Action: The Emotions in Seventeenth-Century Philosophy.* Oxford: Clarendon Press.

Jammer, Max. 1999 (1957). *Concepts of Force: A Study in the Foundations of Dynamics.* Mineola: Dover Publications.

Jaquet, Chantal. 2004. *L'unité du corps et de l'esprit. Affects, actions et passions chez Spinoza.* Paris: Presses Universitaires de France.

2005. *Les expressions de la puissance d'agir chez Spinoza.* Paris: Publications de la Sorbonne.

Jarrett, Charles. 1977. 'The Concepts of Substance and Mode in Spinoza'. *Philosophia* 7 (1): 83–105.

1999. 'Teleology and Spinoza's Doctrine of Final Causes'. In *Desire and Affect: Spinoza as Psychologist. Papers Presented at The Third Jerusalem Conference (*Ethica *III)*, ed. Yirmiyahu Yovel, pp. 3–23. New York: Little Room Press.

2009. 'Spinoza on Necessity'. In *The Cambridge Companion to Spinoza's* Ethics, ed. Olli Koistinen, pp. 118–39. Cambridge University Press.

Jesseph, Doug. 2006. 'Hobbesian Mechanics'. In *Oxford Studies in Early Modern Philosophy, Volume III*, ed. Daniel Garber and Steven Nadler, pp. 119–52. Oxford: Clarendon Press.

Joachim, Harold H. 1993 (1940). *Spinoza's* Tractatus de Intellectus Emendatione*: A Commentary.* Bristol: Thoemmes Press.

Jonas, Hans. 1965. 'Spinoza and the Theory of Organism'. *Journal of the History of Philosophy* 3 (1): 43–57.

Kenny, Anthony. 1970. 'The Cartesian Circle and the Eternal Truths'. *The Journal of Philosophy* 67 (19): 685–700.

2009 (1958). *Descartes: A Study of His Philosophy*. South Bend: St. Augustine's Press.

Koistinen, Olli. 1998. 'On the Consistency of Spinoza's Modal Theory'. *The Southern Journal of Philosophy* 36 (1): 61–80.

2002. 'Causation in Spinoza'. In *Spinoza: Metaphysical Themes*, ed. Olli Koistinen and John Biro, pp. 60–72. Oxford University Press.

2009a. 'Spinoza on Action'. In *The Cambridge Companion to Spinoza's* Ethics, ed. Olli Koistinen, pp. 167–87. Cambridge University Press.

2009b. 'Spinoza's Eternal Self'. In *Topics in Early Modern Philosophy of Mind*, ed. Jon Miller, pp. 151–69. Dordrecht: Springer.

[Forthcoming$_a$] 'Spinoza on Mind'. In *Oxford Handbook of Spinoza*, ed. Michael Della Rocca. Oxford University Press.

[Forthcoming$_b$]. 'Spinoza's Ethics'. In *Great Works of Ethics: Twelve Acts of Reading*, ed. Margit Sutrop and Kadri Simm. Tallinn: Estonian Language Foundation.

Kristeller, Paul Oskar. 1984. 'Stoic and Neoplatonic Sources of Spinoza's *Ethics*'. *History of European Ideas* 5 (1): 1–15.

Lachterman, David R. 1978. 'The Physics of Spinoza's *Ethics*'. In *Spinoza: New Perspectives*, ed. Robert W. Shahan and J. I. Biro, pp. 71–111. Norman: University of Oklahoma Press.

Lærke, Mogens. 2009. 'Immanence et extériorité absolue. Sur la théorie de la causalité et l'ontologie de la puissance de Spinoza'. *Revue philosophique de la France et de l'étranger* 134 (2): 169–90.

[Forthcoming]. 'Spinoza's Cosmological Argument in the *Ethics*'. *Journal of the History of Philosophy*.

Lazzeri, Christian. 1999. 'Spinoza: le bien, l'utile et la raison'. In *Spinoza: puissance et impuissance de la raison*, ed. Christian Lazzeri, pp. 9–38. Paris: Presses Universitaires de France.

Lear, Jonathan. 1982. 'Aristotle's Philosophy of Mathematics'. *The Philosophical Review* 91 (2): 161–92.

LeBuffe, Michael. 2009. 'The Anatomy of the Passions'. In *The Cambridge Companion to Spinoza's* Ethics, ed. Olli Koistinen, pp. 188–222. Cambridge University Press.

2010. *From Bondage to Freedom: Spinoza on Human Excellence*. Oxford University Press.

Lennon, Thomas M. 2005. 'The Rationalist Conception of Substance'. In *A Companion to Rationalism*, ed. Alan Nelson, pp. 12–30. Malden: Blackwell.

Lennox, James G. 1976. 'The Causality of Finite Modes in Spinoza's *Ethics*'. *The Canadian Journal of Philosophy* 6 (3): 479–500.

Lin, Martin. 2004. 'Spinoza's Metaphysics of Desire: The Demonstration of IIIP6'. *Archiv für Geschichte der Philosophie* 86 (1): 21–55.

2005. 'Memory and Personal Identity in Spinoza'. *The Canadian Journal of Philosophy* 35 (2): 243–68.

2006. 'Teleology and Human Action in Spinoza'. *The Philosophical Review* 115 (3): 317–54.

Løkke, Håvard. 2009. 'The Active Principle in Stoic Philosophy'. In *The World as Active Power: Studies in the History of European Reason*, ed. Juhani Pietarinen and Valtteri Viljanen, pp. 49–70. Leiden: E. J. Brill.

LoLordo, Antonia. 2005. 'The Activity of Matter in Gassendi's Physics'. In *Oxford Studies in Early Modern Philosophy, Volume II*, ed. Daniel Garber and Steven Nadler, pp. 75–103. Oxford: Clarendon Press.

Long, A. A. 2003. 'Stoicism in the Philosophical Tradition: Spinoza, Lipsius, Butler'. In *The Cambridge Companion to the Stoics*, ed. Brad Inwood, pp. 365–92. Cambridge University Press.

Macherey, Pierre. 1998. *Introduction à l'*Éthique *de Spinoza. La première partie – La nature des choses*. Paris: Presses Universitaires de France.

Mancosu, Paolo. 1996. *Philosophy of Mathematics and Mathematical Practice in the Seventeenth Century*. New York: Oxford University Press.

Manning, Richard N. 2002. 'Spinoza, Thoughtful Teleology, and the Causal Significance of Content'. In *Spinoza: Metaphysical Themes*, ed. Olli Koistinen and John Biro, pp. 182–209. Oxford University Press.

Marshall, Eugene. 2008. 'Adequacy and Innateness in Spinoza'. In *Oxford Studies in Early Modern Philosophy, Volume IV*, ed. Daniel Garber and Steven Nadler, pp. 51–88. Oxford: Clarendon Press.

Martin, Christopher P. 2008. 'The Framework of Essences in Spinoza's *Ethics*'. *British Journal for the History of Philosophy* 16 (3): 489–509.

Matheron, Alexandre. 1971. *Le Christ et le salut des ignorants chez Spinoza*. Paris: Aubier-Montaigne.

1988 (1969). *Individu et communauté chez Spinoza*. Paris: Les Éditions de Minuit.

1990. 'Le problème de l'évolution de Spinoza du *Traité théologico-politique* au *Traité politique*'. In *Spinoza: Issues and Directions. The Proceedings of the Chicago Spinoza Conference*, ed. Edwin Curley and Pierre-François Moreau, pp. 258–70. Leiden: E. J. Brill.

1991a. 'Essence, Existence and Power in *Ethics I*: The Foundations of Proposition 16'. In *God and Nature: Spinoza's Metaphysics*, ed. Yirmiyahu Yovel, pp. 23–34. Leiden: E. J. Brill.

1991b. 'Physique et ontologie chez Spinoza: l'énigmatique réponse à Tschirnhaus'. *Cahiers Spinoza* 6: 83–109.

Matson, Wallace. 1977. 'Death and Destruction in Spinoza's *Ethics*'. *Inquiry* 20: 403–17.

Mercer, Christia. 2001. *Leibniz's Metaphysics: Its Origins and Development*. Cambridge University Press.

Miller, Jon. 2009. 'Spinoza and the Stoics on Substance Monism'. In *The Cambridge Companion to Spinoza's* Ethics, ed. Olli Koistinen, pp. 99–117. Cambridge University Press.

Molnar, George. 2003. *Powers: A Study in Metaphysics*, ed. Stephen Mumford. Oxford University Press.

Moreau, Pierre-François. 1975. *Spinoza*. Paris: Seuil.

Nadler, Steven. 1998. 'Doctrines of Explanation in Late Scholasticism and in the Mechanical Philosophy'. In *The Cambridge History of Seventeenth-Century Philosophy, Volume 1*, ed. Daniel Garber and Michael Ayers, pp. 513–52. Cambridge University Press.

2006. *Spinoza's* Ethics: *An Introduction*. Cambridge: Cambridge University Press.

2008. '"Whatever Is, Is in God": Substance and Things in Spinoza's Metaphysics'. In *Interpreting Spinoza: Critical Essays*, ed. Charlie Huenemann, pp. 53–70. Cambridge University Press.

Negri, Antonio. 1991 (1981). *The Savage Anomaly: The Power of Spinoza's Metaphysics and Politics*. Trans. Michael Hardt. Minneapolis: University of Minnesota Press.

Nelson, Alan. 2005. 'The Rationalist Impulse'. In *A Companion to Rationalism*, ed. Alan Nelson, pp. 3–11. Malden: Blackwell.

Nolan, Lawrence. 1997. 'The Ontological Status of Cartesian Natures'. *Pacific Philosophical Quarterly* 78 (2): 169–94.

Oittinen, Vesa. 1994. *Spinozistische Dialektik. Die Spinoza-Lektüre des französischen Strukturalismus und Poststrukturalismus*. Frankfurt am Main: Peter Lang.

O'Neill, Eileen. 1993. '*Influxus Physicus*'. In *Causation in Early Modern Philosophy: Cartesianism, Occasionalism, and Preestablished Harmony*, ed. Steven Nadler, pp. 27–55. University Park: Pennsylvania State University Press.

Parkinson, G. H. R. 1971. 'Spinoza on the Power and Freedom of Man'. *Monist* 55 (4): 527–53.

1981. 'Spinoza's Concept of the Rational Act'. *Studia Leibnitiana Supplementa* 20: 1–19.

Pasnau, Robert. 2001. 'Intentionality and Final Causes'. In *Ancient and Medieval Theories of Intentionality*, ed. Dominik Perler, pp. 301–23. Leiden: E. J. Brill.

2004. 'Form, Substance, and Mechanism'. *The Philosophical Review* 113 (1): 31–88.

Pessin, Andrew. 2003. 'Descartes's Nomic Concurrentism: Finite Causation and Divine Concurrence'. *Journal of the History of Philosophy* 41 (1): 25–49.

Pietarinen, Juhani. 2000. 'The Rationality of Desires in Spinoza's Ethics'. In *Rationality and Irrationality (Contributions of the Austrian Ludwig Wittgenstein Society 8)*, ed. Berit Brogaard, pp. 67–72. Kirchberg: Austrian L. Wittgenstein Society.

2001. 'Conatus as Active Power in Hobbes'. *Hobbes Studies* 14 (1): 71–82.

2003. 'Spinoza on Causal Explanation of Action'. In *Realism in Action: Essays in the Philosophy of the Social Sciences*, ed. Matti Sintonen, Petri Ylikoski, and Kaarlo Miller, pp. 137–54. Dordrecht: Kluwer.

2009a. 'Causal Power in Descartes' Mind-Body Union'. In *The World as Active Power: Studies in the History of European Reason*, ed. Juhani Pietarinen and Valtteri Viljanen, pp. 131–62. Leiden: E. J. Brill.

2009b. 'Motion and Reason: Hobbes' Difficulties with the Idea of Active Power'. In *The World as Active Power: Studies in the History of European Reason*, ed. Juhani Pietarinen and Valtteri Viljanen, pp. 185–212. Leiden: E. J. Brill.

Ramond, Charles. 1995. *Qualité et quantité dans la philosophie de Spinoza*. Paris: Presses Universitaires de France.

1999. 'Impuissance relative et puissance absolue de la raison chez Spinoza'. In *Spinoza: puissance et impuissance de la raison*, ed. Christian Lazzeri, pp. 63–91. Paris: Presses Universitaires de France.

2007. *Dictionnaire Spinoza*. Paris: Ellipses.

Rice, Lee C. 1971. 'Spinoza on Individuation'. *The Monist* 55 (4): 640–59.

1985. 'Spinoza, Bennett, and Teleology'. *The Southern Journal of Philosophy* 23 (2): 241–53.

1992. 'La causalité adéquate chez Spinoza'. *Philosophiques* 19 (1): 45–59.

1998. 'Spinoza and Highway Robbery'. *Archiv für Geschichte der Philosophie* 80 (2): 211–18.

1999. 'Action in Spinoza's Account of Affectivity'. In *Desire and Affect: Spinoza as Psychologist. Papers Presented at the Third Jerusalem Conference (*Ethica *III)*, ed. Yirmiyahu Yovel, pp. 155–68. New York: Little Room Press.

Rizk, Hadi. 2006. *Comprendre Spinoza*. Paris: Armand Colin.

Rousset, Bernard. 1994. 'Les implications de l'identité spinoziste de l'être et de la puissance'. In *Spinoza: puissance et ontology*, ed. Myriam Revault d'Allonnes and Hadi Rizk, pp. 11–24. Paris: Éditions Kimé.

Sachs, Mendel. 1976. 'Maimonides, Spinoza, and the Field Concept in Physics'. *Journal of the History of Ideas* 37 (1): 125–31.

Scala, André. 1994. 'Puissance et définition. La proposition 16 du Livre I de l'*Éthique*'. In *Spinoza: puissance et ontology*, ed. Myriam Revault d'Allonnes and Hadi Rizk, pp. 25–38. Paris: Éditions Kimé.

Schmaltz, Tad. 1991. 'Platonism and Descartes' View of Immutable Essences'. *Archiv für Geschichte der Philosophie* 73 (2): 129–70.

2008. *Descartes on Causation*. Oxford University Press.

Schmidt, Andreas. 2009. *Göttliche Gedanken. Zur Metaphysik der Erkenntnis bei Descartes, Malebranche, Spinoza und Leibniz*. Frankfurt am Main: Vittorio Klostermann.

Schrijvers, Michael. 1999. 'The *Conatus* and the Mutual Relationship between Active and Passive Affects in Spinoza'. In *Desire and Affect: Spinoza as Psychologist. Papers Presented at the Third Jerusalem Conference (*Ethica *III)*, ed. Yirmiyahu Yovel, pp. 63–80. New York: Little Room Press.

Secada, Jorge. 2000. *Cartesian Metaphysics: The Late Scholastic Origins of Modern Philosophy*. Cambridge University Press.

Sévérac, Pascal. 1998. 'Passivité et désir d'activité chez Spinoza'. In *Spinoza et les affects*, ed. Fabienne Brugère and Pierre-François Moreau, pp. 39–54. Paris: Presses de l'Université de Paris-Sorbonne.

2005. *Le devenir actif chez Spinoza*. Paris: Honoré Champion.

Spindler, Fredrika. 2005. *Philosophie de la puissance et détermination de l'homme chez Spinoza et chez Nietzsche*. Göteborg: Glänta Produktion.

Steenbakkers, Piet. 2009a. 'The Textual History of Spinoza's *Ethics*'. In *The Cambridge Companion to Spinoza's* Ethics, ed. Olli Koistinen, pp. 26–41. Cambridge University Press.

2009b. 'The Geometrical Order in the *Ethics*'. In *The Cambridge Companion to Spinoza's* Ethics, ed. Olli Koistinen, pp. 42–55. Cambridge University Press.

Steinberg, Diane. 2000. *On Spinoza*. Belmont, CA: Wadsworth.

2005. 'Belief, Affirmation, and the Doctrine of *Conatus* in Spinoza'. *The Southern Journal of Philosophy* 43 (1): 147–58.

Van Zandt, Joe D. 1986. '*Res extensa* and the Space-Time Continuum'. In *Spinoza and the Sciences*, ed. Marjorie Grene and Debra Nails, pp. 249–66. Dordrecht: Reidel.

Viljanen, Valtteri. 2009a. 'Schopenhauer's Twofold Dynamism'. In *The World as Active Power: Studies in the History of European Reason*, ed. Juhani Pietarinen and Valtteri Viljanen, pp. 305–29. Leiden: E. J. Brill.

2009b. 'Spinoza's Actualist Model of Power'. In *The World as Active Power: Studies in the History of European Reason*, ed. Juhani Pietarinen and Valtteri Viljanen, pp. 213–28. Leiden: E. J. Brill.

2009c. 'Spinoza's Ontology'. In *The Cambridge Companion to Spinoza's* Ethics, ed. Olli Koistinen, pp. 56–78. Cambridge University Press.

2010. 'Causal Efficacy of Representational Content in Spinoza'. *History of Philosophy Quarterly* 27 (1): 17–34.

Von Dunin Borkowski, Stanislaus. 1933. 'Die Physik Spinozas'. In *Septimana Spinozana*, pp. 85–101. The Hague: Martinus Nijhoff.

Watt, A. J. 1972. 'The Causality of God in Spinoza's Philosophy'. *The Canadian Journal of Philosophy* 2 (2): 171–89.

Wilson, Margaret D. 1978. *Descartes*. London: Routledge.

1983. 'Infinite Understanding, *Scientia Intuitiva*, and *Ethics* 1.16'. *Midwest Studies in Philosophy* 8 (1): 181–91.

1991. 'Spinoza's Causal Axiom (*Ethics* 1, Axiom 4)'. In *God and Nature: Spinoza's Metaphysics*, ed. Yirmiyahu Yovel, pp. 133–60. Leiden: E. J. Brill.

1996. 'Spinoza's Theory of Knowledge'. In *The Cambridge Companion to Spinoza*, ed. Don Garrett, pp. 89–141. Cambridge University Press.

Wilson, Mark. 1998. 'Field Theory, Classical'. In *Routledge Encyclopedia of Philosophy, Volume III*, ed. Edward Craig, pp. 668–70. London: Routledge.

Wolf, A. 1974 (1927). 'Spinoza's Conception of the Attributes of Substance'. In *Studies in Spinoza: Critical and Interpretive Essays*, ed. S. Paul Kashap, pp. 16–27. Berkeley: University of California Press.

Wolfson, Harry Austryn. 1961 (1934). *The Philosophy of Spinoza. Two Volumes in One*. Cleveland and New York: The World Publishing Company.

Woodfield, Andrew. 1976. *Teleology*. Cambridge University Press.

Youpa, Andrew. 2003. 'Spinozistic Self-Preservation'. *The Southern Journal of Philosophy* 41 (3): 477–90.

Index

For EU product safety concerns, contact us at Calle de José Abascal, 56–1°,
28003 Madrid, Spain or eugpsr@cambridge.org.

www.ingramcontent.com/pod-product-compliance
Lightning Source LLC
LaVergne TN
LVHW011521200926
848484LV00001B/96

* 9 7 8 1 1 0 7 0 0 7 8 0 2 *